Telecommunications

Publications of the

Münchner Kreis

Supranational Association for Communications Research

Volume 20

Global Players in Telecommunications

Proceedings of a Congress
Held in Munich, April 20/21, 1994

Editor: E. Witte

Springer-Verlag
Berlin Heidelberg New York
London Paris Tokyo
Hong Kong Barcelona Budapest

Münchner Kreis
Übernationale Vereinigung für Kommunikationsforschung
Supranational Association for Communications Research
Tal 16, D-80331 München, Telefon: (089) 22 32 38

Chairman of the Congress:

Prof. Dr. Dres. h. c. Eberhard Witte
Institut für Organisation
der Universität München
Ludwigstraße 28, Rgb.
D-80539 München

ISBN 3-540-58488-9 Springer-Verlag Berlin Heidelberg New York
ISBN 0-387-58488-9 Springer-Verlag New York Berlin Heidelberg

Satz: Reproduktionsfertige Vorlagen der Autoren
SPIN: 10470061 62/3020 - 5 4 3 2 1 0 - Gedruckt auf säurefreiem Papier.

Contents

Opening Address

Eberhard Witte

Telecommunications enable people separated by great distances to exchange information -- people who otherwise could not keep in contact without the help of technology. In this sense, the fundamental idea behind telecommunications has always been focused on long-distance systems -- systems which ultimately spanned the entire globe. The first telegraph line between England and India, built in 1867, could be called the first true global communications system. The geographical aspect of telecommunications, then, is nothing new to discuss at this congress. The worldwide switching system linking calls among more than 600 million people with phones is actually the largest automated machine in the world. This system bridges all continents and seas of the earth.

The traditional systems used to achieve this miracle of technology were not internationally organized from the outset, but very much nationally oriented. In the past, the world's telecommunication systems were distinguished by three features:

- In all countries throughout the world -- even in the United States with its privately operated telephone systems -- telecommunications was considered to be a public service for the general population. In most countries, the government organized telecommunications as a public authority -- as a postal, telephone and telegraph system combined in one organization.

- Universal services stood in the foreground: telegraphy, telephony, followed in later years by text transmissions and simple data exchanges. Nationwide service, uniform tariffs and the obligation to operate for the common good were declared as the principles behind government control of telecommunications.

- These monopolies dominated the development of telecommunications in every country around the world up until 1982. Telecommunication networks were frequently viewed as natural monopolies; competition was not only undesired, but considered to be impossible.

With the beginning of deregulation in the United States -- a trend that quickly spread to Britain, Japan and other countries -- the picture rapidly changed:

- State-owned telecommunications administrations were privatized step by step.

- Universal services were supplemented with value-added services, with special offers targeted at specific user groups. At the same time, there was a parallel development of various network systems (cable networks, terrestrial radio networks, satellite networks).

- Territorial monopolies were gradually loosened up and ultimately eliminated completely. Competition first arose on the periphery of the telecommunications industry (mobile radio, paging systems, satellite services), and gradually spread to the very heart of telecommunications -- telephony.

These developments were preceded by the accelerating internationalization of major businesses. Companies in the automobile, chemical and engineering industries, as well as banks and insurance firms spread their activities around the world. They no longer wanted to deal exclusively with their traditional national carrier, who in turn passed on their communication needs to other national carriers. Instead, they increasingly sought a single partner who would look after them worldwide and guarantee them the same tailored services, the same software, and standardized terminals wherever they were located.

This new diversity of telecommunication suppliers and complete entrepreneurial freedom spawned by deregulation has called the old system of national, limited telecommunication systems into question. The new vendors on the market as well as the former state-owned operators that are now privatized no longer view national borders as a boundary for their business activities: they are moving aggressively into foreign markets.

Companies that have abandoned their original concentration on national markets to move into the international arena can be called global players. Their strategies for building up international business take various forms:

- The simplest way to move into the international market is by acquiring another national telecommunications company. This has been the tactic used by European companies throughout South America. If this policy of company merger and acquisition would be carried to the extreme, a single global player would ultimately emerge. The resulting telecommunications company would then operate throughout the world and have a global rather than a national monopoly. This development is undesirable both politically and economically.

- One alternative is for a national carrier to found a subsidiary in another country. Once this subsidiary obtains an operating license in the host country, it automatically creates a competitive situation in a national market formerly dominated by the domestic carrier. This is the type of market penetration currently being employed by mobile radio, paging and data communication companies.

- The most commonly used tactic is the reciprocal exchange of capital between public telephone operators (PTOs). Usually the participants found a joint venture, which then operates in a number of countries. These joint ventures continue to have a national home market, yet achieve geographical diversity by spreading into foreign markets as well.

- One future-oriented alternative is to form strategic alliances. In these cases, there is no need to exchange equity capital or found new subsidiaries outside the home market. Rather, the national PTOs cooperate in various markets around the world by signing agreements and coordinating their long-term global strategies. In this manner they can offer global services on global networks, and remain competitive with other strategic alliances pursuing the same objectives.

The services offered in this global game include:

- national and transnational telecommunication services,

- basic services and enhanced services, particularly innovative services.

The strategy is primarily aimed at providing enhanced services on the international market. Networks are usually leased and seldom built by the company itself. Terminals and peripherals are procured on the open market, and are generally compatible. Vertical integration -- particularly the linking of telecommunication services and the production of telecommunications hardware -- is undesirable, yet still exists and has re-emerged in certain countries.

This development has become so obscured and complex in recent years, that the Congress of the Munich Circle has decided to attempt to obtain a clear overview of the current international situation. This project is of particular interest to the German telecommunications industry, since the postal and telecommunications reform now being carried out in various stages in Germany will ultimately create a completely open and competitive market that will attract the global players. On the other hand, German companies want to grasp opportunities to participate in the international market and also form their own strategic alliances.

It has been especially fortunate that the CEOs of the world's largest telecommunication companies have agreed to participate in this congress and present their respective global strategies.

As host of this Congress, I wish to thank all our guests who are prepared to openly talk about their company strategies and discuss the major strategic alternatives of the global players.

The Implications of Global Telecommunications Systems for the ITU

Pekka Tarjanne

Mr Chairman,
Ladies and Gentlemen,

It gives me great pleasure to address this distinguished conference on the subject of global telecommunications. The very *raison d'être* of the International Telecommunication Union is to be "global". Indeed, the purposes of the Union, as stated in the Geneva Constitution, include the provisions to "maintain and extend international co-operation ... " to "promote the extension of the benefits of the new telecommunication technologies to all the world's inhabitants" and to "promote, at the international level, the adoption of a broader approach to the issues of telecommunications in the global information economy and society ...". So the ITU has always been global in scope. But whereas our Members have traditionally represented sovereign national interests, they now increasingly also represent global interests. And whereas most of the services they provide were originally intended primarily for domestic customers, they are now increasingly targeted at foreign and multinational clients. In this paper, I wish to examine the nature of globalisation in the telecommunication sector and to consider what implications it has for the work of the ITU and its Members.

Telecommunications and globalisation

The idea that we are living in a *global* or *borderless* world is relatively new. The meaning we now attach so easily to the word *global* was unknown 100 years ago when the world was still in the process of being divided up into independent sovereign nation states. However, the dramatic events which have taken place in the twentieth century -- including the development of global transportation and telecommunication systems, and the rise of global products, markets, and corporations -- have convinced many people that we are indeed living in a new historical era in which the economic, social, cultural and political structures that shaped relations between people over the past two centuries will be transformed. Though accounts differ as to why this is so, most emphasise the central role played by developments in telecommunications, computer and information technology.

This historical change is often described as the passage from an *industrial* to an *information* era.

Today, the term *global village* conjures up the picture of a vast, worldwide communications network of interconnected satellite and land and undersea cable links enabling the rapid transfer of growing voice, data and image traffic. But if truth be told, the global village is still very much a dream because the majority of the world's population does not have the resources or the infrastructure to access communication networks (Figure 1). Yet the technological capability to reach distant places via microwave, cable or satellite *is* a reality and the expanding web of communications by land, sea and air, spurred by the dramatic growth in international voice, data, and image traffic brings us ever closer to a true village.

Figure 1: Unequal shares of the global network: Telephone main lines per 100 inhabitants, 1992

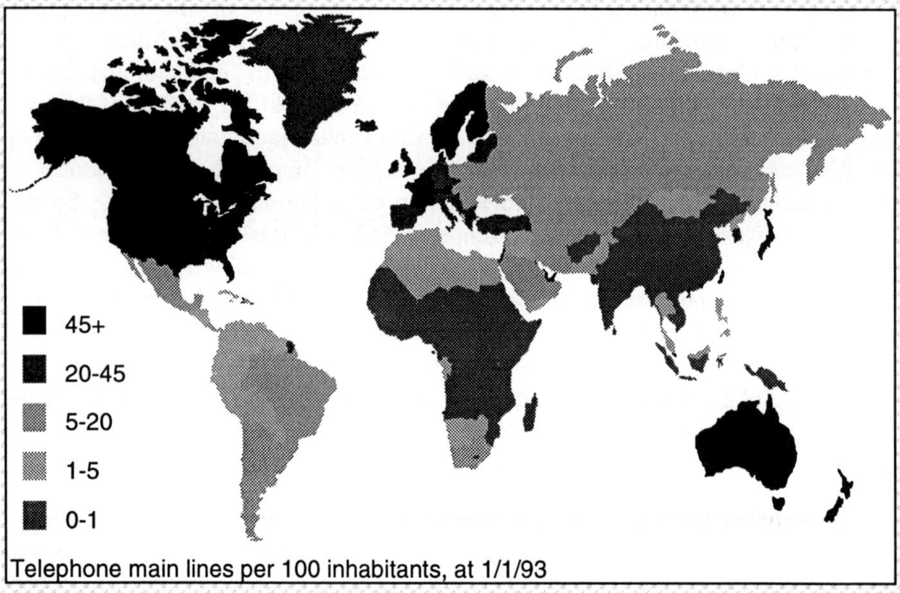

Source: ITU (1994) "World Telecommunication Development Report".

Perhaps the clearest evidence for globalisation is found in the continuing increase in the volume of **international telecommunications traffic** both in absolute terms and as a percentage of total traffic. International traffic is the fastest growing part of the telecommunications business and also the most profitable. International telephone traffic has been growing at some 17 per cent per year over the last decade, fuelled by growth in international trade, travel and immigration and a general expansion of the global economy. In 1992, the inhabitants of the globe generated more than 40 billion minutes of international telephone traffic (Figure 2). This equates to around seven minutes per person or 70 minutes per subscriber. Usage per telephone has more than doubled in the last ten years.

Figure 2: The lines are busy

International outgoing telephone traffic, 1983-92, and by region, 1992

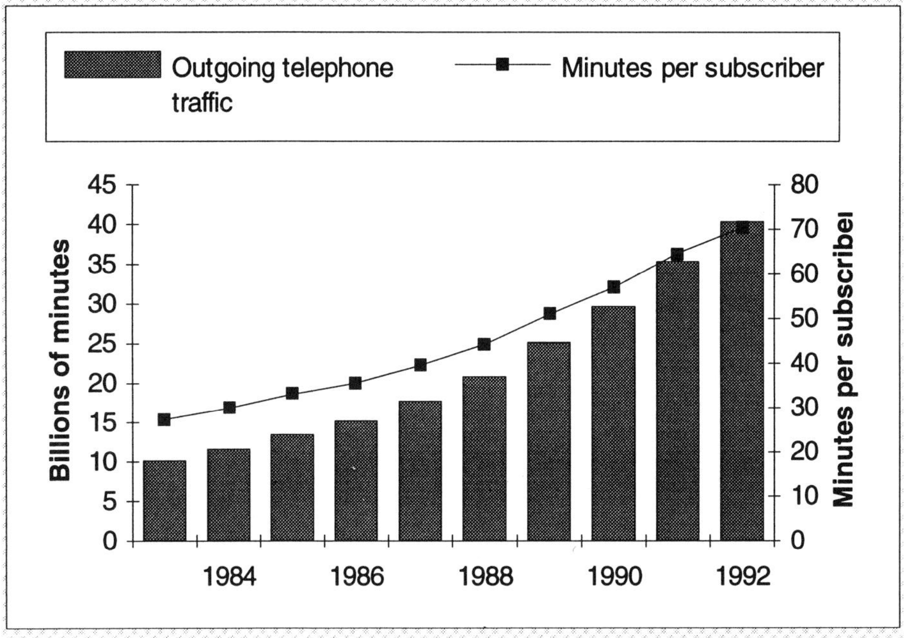

Source: ITU/TeleGeography Inc (forthcoming, 1994) "Direction of Traffic: World Telephone Flows, 1983-92".

A second measure of globalisation is the extent to which the activities of individual manufacturers and service providers cross national boundaries. The **corporate strategies** of today's global players, many of whom are represented at this conference, are increasingly taking them into foreign markets both as an aggressive move, to develop new business opportunities, and as a defensive move, because of incursions into their home market or due to declining growth in mature networks. This is occurring through mergers and acquisitions, through foreign direct investment and the establishment of subsidiaries, through the formation of consortia and alliances, and through an increase in export sales as a percentage of total sales. The top 16 telecommunications manufacturers, with combined sales of almost US$ 100 billion, gained more than 50 per cent of their sales from outside their domestic markets. For the public telecommunication operators, this share is much lower but is growing nonetheless. International telecommunication services generated just under 15 per cent of the US$415 market for public telecommunication services in 1992.

Finally, perhaps the best measure of globalisation is the ability of companies to provide **end-to-end service to users**, wherever they may be located. Many countries still retain rules which limit the extent of service provision by foreign-owned companies and oblige foreign companies to interconnect with a monopolistic public telecommunications operators. However, the traditional model of jointly-provided international services is slowly breaking down and a growing number of countries now permit competition in the provision of both telecommunication services and infrastructures. Experience with liberalisation has shown that, once competition has been introduced, it is difficult to place arbitrary limits on the extent to which new market players can compete with the monopoly service provider: mobile communications companies want to construct their own high-speed fixed-link networks between cities; value-added network suppliers want to be able to provide services direct to client sites; private network owners want to be able to resell spare capacity. In all cases, the new market entrants will sooner or later want to provide international services independently from the incumbent carriers. Thus globalisation almost always brings new pressures on national competition policy.

Globalisation and market structure

It has become the conventional wisdom within the telecommunication services industry that, over time, there will be a *shake-out* of the number of major players. According to this view, a process of market rationalisation -- achieved through acquisitions, mergers and global alliances -- will create a two-tier market structure. At the top level will be a handful of truly global players offering a full range of telecommunication services. They will draw upon the services of a much larger second tier of operators with strengths either in one national market or in one particular technology (e.g., mobile, satellite). This scenario is easy to believe. Looking at other industries, say the automobile, soft drinks or even a closely related industry such as telecommunication switch manufacturing, it can be seen how market rationalisation has reduced the number of players possessing global brands to single figures.

In other industries which have traditionally been highly regulated -- airlines, postal services or railways -- national companies still dominate their home market. A second possible scenario is one in which the foreign ventures of public telecommunication operators increase and a number of smaller countries invite in foreign operators, but the activity remains at the margin and most large countries retain national carriers. This second scenario would represent a continuity in the history of the industry which has hitherto been based on national sovereignty and international co-operation.

However, there is also a third, and equally plausible scenario for the future evolution of the telecommunication services industry. Here the driving force is neither economies of scope and scale nor national sovereignty, rather it is rapid technological change. It can be argued that the development of new communication technologies, particularly radio-based but also interactive multimedia services carried over high bandwidth cables, is introducing to the telecommunications market a set of new players without the history or the installed base of traditional public telecommunication operators. For instance, companies such as Vodafone (UK) and Millicom (registered in Luxembourg), are pursuing aggressive foreign investment strategies as new market players without an existing fixed-link installed base. Under this scenario, traditional public telecommunication operators will continue to grow, but theirs would be a diminished share of a much larger, and far more fragmented total communications market based on different technological platforms such as copper, or fibre, or radio or satellite. Competition will be expressed as much in concepts such as value for money, quality of service and customisation to individual clients as in technical differences between service offerings. Thus the overall level of concentration in the industry may decline, at least in the short term. Global players may emerge, but only in specific market segments.

These three scenarios are summarised in Figure 3. In practice they are not incompatible and all three might co-exist in different parts of the industry or in different regions of the world. All three scenarios would provide logical outcomes from the processes of globalisation described earlier and it would be possible to make a transition from one stage to another. However, each scenario would have different implications for the role of the ITU. These are discussed below.

Global telecommunication systems: Do we need global regulation?

Of the three scenarios listed above, the ITU in its current form is optimised towards the "national sovereignty" scenario (2). ITU could cope well with the "fragmented markets" scenario (3), in that it already has a decentralised, federal structure in which each of the three sectors has a high degree of autonomy. But the "global players" scenario (1) would require a substantial rethink of the ITU's mission and strategic direction. Some would argue that a legitimate response to the globalisation of the telecommunications industry is to move towards a system of global regulation. There are several powerful arguments which would appear to favour this viewpoint:

Figure 3: Alternative scenarios for the future evolution of the telecommunication services industry

Scenario	*Characteristics*	*Assumptions*	*Similar industries*
1. Market rationalisation combined with globalisation.	A handful of global telecommunication services companies come to dominate the industry. They gain market position through mergers and acquisitions and form alliances with a secondary tier of smaller players specialised on one country or one market sector.	Economies of scale and scope are main driving factor meaning that only a few players can achieve sufficient scale to offer a full range of services across the globe. Ownership of global cable network and/or satellite system is vital.	Telecom switching manufacture. Automobiles. Chemicals. Soft drinks.
2. Continuing *status quo* with national carriers dominant in home market.	Most of the major countries retain a national carrier with majority ownership. Some competition and foreign ownership in marginal areas such as mobile, satellite or VANs.	National sovereignty is the dominant factor based on concerns of national security and industrial policy.	The railway industry. The airline industry. Postal services.
3. Multiple and diverse players entering fragmented telecom market.	High levels of market entry in fields such as mobile communications, data communications and multimedia to challenge fixed-link PTOs. Competition between different technologies. Fragmentation of telecommunications industry into a series of technology-specific and regional markets.	Rapid technological change reduces economies of scale permitting firms in other industries (computers, entertainment, etc.) to invest in telecommunications. Traditional PTOs hold declining share of growing market and face competition from companies with alternative technology platforms (radio, satellite, cable TV).	Industries experiencing rapid technological change such as computer industry in the early 1980s, pharmaceutic-als industry, biotechnology

Note: PTO = Public Telecommunication Operator.
Source: ITU (1994) World Telecommunication Development Report.

- The first argument is based on the sheer **size of the public telecommunication industry**. In 1992, the public telecommunications industry, services and equipment combined, was worth some US$535 billion and contributed around 2.4 per cent of global Gross Domestic Product (GDP). Some 42 public telecommunications operators now have annual sales in excess of US$1.5 billion. Of these companies, the majority are privately-owned, or are expected to be privatised in the near future. By contrast, more than 50 of the ITU's 183 Member countries have a GDP of less than US$1.5 billion. Many national governments do not have the ability or the resources to regulate the new generation of global telecommunication players. Even fewer countries can systematically ensure that regulations are enforced.

- Second, many of the problems that face telecommunication policy-makers today are fundamentally global, rather than national or regional, in nature. By the end of this decade, we could see the development of **global mobile satellite services** offering personal communication services to handheld receivers anywhere in the world. The promise of this technology is that it makes any locality potentially accessible to the global telecommunication network. But at the same time it raises issues of global co-ordination, co-operation and competition that have not previously been faced. Several national administrations are currently competing to establish pre-emptive regulatory structures which would license carriers for the provision of services which are essentially international in character;

- Other **resource allocation** issues also demand a global approach. Many new services being developed need access to the radio frequency spectrum and the global numbering plan. But traditional *first come, first served* allocation mechanisms do not lend themselves to competitive markets or those in which scarcity is an important factor. Often, market access in telecommunications is equivalent to access to resources or to public networks, and the fact that companies wishing to create global services have to make separate agreements in each of the countries in which they wish to operate imposes substantial extra costs. This can deter them from entering smaller or more marginal markets which is to the detriment of the goal of universality;

- The **reform of the accounting rate process** needs to be tackled on a multi-lateral basis Accounting rate payments provide an important source of income to some of the developing countries but they also tend to keep telephone charges to end users high, and they can act as a disincentive to economic efficiency. The challenge here is to reform the system while still maintaining the vital flow of investment funds to assist with network development.

- Finally, as noted above, the liberalisation of international trade in services, the rise of foreign direct investment and the forging of inter-firm alliances has created a new set of **global players** in the international telecommunications market. What forms of global competition policy would be necessary to ensure that global corporations continue to work for the benefit of their customers as well as their shareholders?

At face value, there would appear to be a strong case for a new, powerful regulatory body that would bring together the activities of national regulatory authorities. But there are problems associated with such an approach:

- **The lack of appropriate public institutions.** The two existing bodies that could fulfil such a role -- the ITU and the WTO -- do not have the appropriate expertise. The ITU's staff profile is heavily biased towards engineering and administration skills and we have relatively little experience in policy-making or regulation, though we are taking steps to remedy this. The World Trade Organisation, the successor to the GATT, is even less able to take on the mantle of becoming a global regulator because of its lack of expertise in this area, though the recent General Agreement on Trade in Services includes a specific annex on telecommunications;

- **The dominance of industrial policy interests.** For much of its history, ITU has been hamstrung in its ability to act by World Wars, by the Cold War and by conflict between first and third world interests. For this reason, the ITU has opted to take technocratic approach, specialising in those areas in which ideology has little to contribute and little to detract, such as technical interface standards or frequency assignment and registration. But, with the emergence of the so-called *triad* of industrial policy interests -- North America, the European Union, Japan -- even these technocratic areas are now cut through by commercial interests. Thus the ITU's ability to act in emerging technologies such as high definition television, asynchronous transmission mode or personal communications mobile satellite systems is seriously impaired. There is no reason to believe that giving the ITU a formal regulatory role would make reconciling these divergent commercial interests any easier;

- **The lack of appropriate regulatory mechanisms.** The regulatory tradition is still relatively young in telecommunications, at least outside North America, and the models which have been established for regulatory procedure can be rapidly overtaken by technological change. The public telecommunications industry has historically been "carrier regulated", in contrast to the "content regulation" of the broadcasting sector. But neither model lends itself well to the international stage. Furthermore, concepts such as universal service, open access, free and fair competition, reciprocal market access or cost-oriented pricing become increasingly elusive when taken out of a particular national context. Also, the ability to regulate effectively has to be backed up by the ability to take sanctions against a particular service provider, such as financial penalties, licence withdrawal or price-caps. It is unlikely that such sanctions could be applied in an international context.

For these reasons and more, it is unlikely that there could be any acceptable legal or political basis for global regulation in the telecommunications sector, even though the arguments in favour of it may seem quite powerful. It would be better therefore to concentrate on what can be done to improve *global co-ordination,* between national and regional regulatory bodies, rather than seeking to impose *global regulation.* Within the context of our strategic planning for the forthcoming ITU Plenipotentiary Conference in Kyoto later this year, the ITU is building on its existing strengths and proposing a number of initiatives which would help to address this issue:

- The ITU should continue to be guided by the goals and principles that are set out in the Constitution. They are the "moral compass" of the ITU. They help to define our role and distinguish it from those of other members of the telecommunications community. First and foremost among these is the goal of fostering and facilitating the global development of telecommunications for the universal benefit of mankind. This is a practical challenge of immense importance to the international community. In the twenty-first century economic, social and cultural development, as well as global peace and security, will depend on good communications. The symbolic functions of the ITU are also very important. Symbolically, the ITU represents the rights and obligations of all members of the global community. It seeks to advance the common interests of mankind in telecommunications through the rule of law, mutual consent and co-operative action. Its actions are based on fundamental respect for the human need to communicate and recognition of the role that telecommunications increasingly plays in sustainable economic, social, cultural and political development.

- In the 1995-99 draft strategic plan for the Union, we have proposed the development of an ITU core competence in the area of telecommunications policy analysis and regulation, and the creation of a forum for Members to exchange information and share ideas on how to tackle the major policy issues of the day. These steps will build on the success of the ITU regulatory colloquia which were established in 1993. Now that the functions of operation and regulation are split in most ITU Member countries, it is important that the ITU provides an appropriate meeting place for regulators as well as telecommunication manufacturers and operators.

- The trends towards globalisation outlined above clearly have important implications for all ITU members, both in the public and in the private sector. To deal with them effectively, new patterns of partnership will have to be developed to unite the interests of government and the private sector in the development of global telecommunications. This will almost certainly require enhanced participation in the affairs of the Union by our growing number of private sector members.

- Against this background, the mission of the ITU should be *"to provide the products and services required by the global telecommunications community that are best provided by an international organisation"*. Carrying out this mission in the changing telecommunications environment will require the ITU to become an open organisation, responsive to the needs of all members of the telecommunications community, client-oriented, entrepreneurial and risk-taking in its behaviour. This will almost certainly entail a re-evaluation of the portfolio of products and services which are offered by the Union to its Members;

In conclusion, it is clear that the telecommunications environment in which we work is changing at an ever more rapid pace. It is sometimes difficult for a 130-year old organisation such as the ITU to keep up with what is new in our industry. But fortunately most of our members are much younger and continue to push us to respond to the changing environment. The central strategic challenge facing the ITU is to adapt an intergovernmental organisation based on national sovereignty to the emerging requirements of the global information society, in which telecommunication is becoming an increasingly important business activity, as well as remaining a key socio-economic infrastructure and an essential public service. ITU does not have the capability or the desire to become a global regulator. But its role as a forum for information exchange and for global co-ordination of national and regional regulatory policy is more important than ever.

Putting Global Costumers First

Sir Iain Vallance

The telecommunications industry is in the throes of a series of tremendous changes. We have heard that sort of statement for years ... but now it is actuallyhappening. Around the world the traditional, national, monopolybased structures are breaking down. At the same time, the long-heralded and much discussed convergence of the telecommunications, computing, publishing, broadcasting, information and entertainment industries is taking place.

Global customers are increasingly demanding global service. There is pressure, resisted by some but welcomed by others, to liberalise markets. Much of this pressure comes from customers anxious to benefit from the services (andcompetitive prices) that such convergence allows - particularly across national boundaries.

There is support for the liberalisation of markets by the business community right across Europe. The majority see telecommunications as the industry they would most like to see liberalised. Research that we commissioned in eight European countries clearly shows that the vast majority of business customers believe that telecommunications liberalisation is a process fundamental to the creation of a single market. They believe it will increase the range, improve the quality and reduce the costs of telecommunications services.

I believe that the telecommunications industry in Europe has so far failed to respond to these expectations. We have failed because we have not reacted quickly enough to match the needs of our customers.

What do global customers want?

They want seamless end-to-end service, region to region, country to country, desk to desk. They want customised billing. One bill, in a currency or currencies of their choosing at a frequency of their choice. They want one point of contact for their telecommunications needs with service available 24 hours a day, seven days a week. These are just some of the services that customers tell us that they want.

What is stopping operators such as BT, France Telecom and Deutsche Bundespost Telekom from providing these services on a global or even pan-European basis? As I have already said it is not customers reticence - it is not technology. The real resistance comes from trade barriers that were erected in an era when national boundaries defined the extent of telecommunications activities. These are no longer relevant and they need to be dismantled quickly to ensure that markets and operators can respond to customer needs.

Incumbent PTT operators are still fearful of such liberalisation. But what have they to fear? Experience within the US and UK shows that everyone benefits; customers, competitors and even the incumbent PTT. I firmly believe that most operators will prosper in liberalised markets.

Experience in the UK helps to make this point. Thanks to liberalisation, British companies have amongst the lowest telecommunications charges in Europe and the widest range of services and suppliers. The UK Government's Telecommunications Act of 1984 and its White Paper "Competition and Choice" in 1991 created an environment in which a wide range of operators could compete head to head with BT. Currently there are over 140 Public Telecom Operators including over 40 cable television companies. Recent licences granted include Telia from Sweden, Telstra from Australia, and Sprint and Worldcom from the US with a further 31 applications being considered. BT, by definition, has lost market share. But what matters is not so much losing percentages of market share than growing the absolute size of the market.

In the area of cellular communications I have no doubt that the explosive growth in these services in the UK would not have happened without the existence of two competing networks, Cellnet and Vodafone, from the outset.

Here competition has had a positive effect of significantly accelerating the introduction of new products and services. For example voice messaging has achieved high levels of penetration; call access can turn the mobile phone into an office extension; and both operators offer a range of tariffs suitable for high or low users, business or personal customers, on a local or national basis.

Fierce competition has also generally stimulated the market. An example here is the impact that the entry of Mercury and US West's One-2-One PCS service has had on an already highly competitive cellular market. One-2-One was launched in September 1993 yet our cellular operator, Cellnet, added more customers in December 1993 than during the whole of 1992!

These simple examples demonstrate quite clearly how openness, competition and relative freedom from regulation can create tremendous opportunities for the "incumbent" operator as well its competitors. With customers in mind, the next, vital move is to translate this national experience into the international arena.

I have said that global customers demand global service. Technology allows them to seek such service from a supplier based in any part of the world. This gives a natural advantage to companies operating in markets that allow and encourage competition. In this respect I think it is vital that Europe looks across the Atlantic to see the impact that the debate about information highways and multimedia opportunities has had in the US.

The discussion in the United States provides excellent clues about the future direction of the market. In particular, "convergence" is encouraging the US administration to rethink the nature and role of regulation. The concept of multimedia services has focused attention on the different types and degrees of regulation, which range from the fairly tight control of the telecoms market, (following the break-up of AT&T in 1984) to the relatively unregulated computing markets.

Fundamental to the US Government's concept of a National Information Infrastructure is the recognition that private sector expertise and funding will play a central role in the creative, and massively competitive development of such a broadband system and associated services. The US Administration has also recognised that this creative energy from the private sector will not be engaged on the scale required unless the regulatory systems are radically overhauled. Convergence compels deregulation - the "information age" is predicated on free competition.

Vice President Gore recently summed up the prevailing view very succinctly when he said that "... this Administration will not let existing regulatory structures impede or distort the evolution of the communications industry".

The Brooks-Dingell, Markey-Fields and other initiatives of the Senate and Congress will open up the local, long distance and international markets to more competition and will provide the necessary environment for private enterprise to flourish, in a framework broadly defined by the Administration.

Looking at this from both a European and global perspective, I think it is important to grasp that what the Vice President first described as a National Information Infrastructure is clearly evolving into an International Information Infrastructure.

PTTs and other operators in the European Union have a vital role to play in promoting the International Information Infrastructure and, consequently, need to reassess the regulatory barriers to progress. And the European Union (and the individual nation states within the Union) must now take the initiative to undertake a similar radical reassessment of their markets and their regulation. The alternative is to fall further and further behind our more far sighted and aggressive international competitors in the US and Japan.

Last July the Council of Telecoms Ministers agreed that public voice services would be liberalised - but not until 1998 - with some member states receiving a period of grace until 2003. I believe this is far, far too slow... especially, as I have said, if enabling US legislation is enacted in 1994. Without a single market for telecommunications and multimedia, Europe will find itself stranded in the twentieth century.

The Delors White Paper and Bangemann High Level Working Group are significant and encouraging contributions to this debate. At least we are now discussing "when" and "how", not "whether" to liberalise.

The White paper has identified the need for a European vision of the information society with the focus on Trans European Networks. The Bangemann High Level Group will make proposals to the European Council of Ministers' summit in Capri in June. And the signs so far are that both its working groups favour much more rapid deregulation and liberalisation of telecommunications markets in Europe. A key element here that will drive costs down is the encouragement of competition in infrastructure. Let's hope that they are not deflected from this enlightened course by the protectionists!

Turning now to BT's response to the challenges of global customer requirements, what have we done to help this process of promoting global liberalised, competitive markets?

First, of all the European players, BT has been the most vociferous in calling for the realisation of the single European Act in the field of telecommunications. It has made us enemies as well as friends.

Second, BT has advocated and supported a wide range of liberalisation initiatives undertaken in our home market, in the UK, by both the Government and the regulator, Oftel.

Finally, BT has established a joint venture company with a partner with impeccable free market credentials, to address the global communications requirements of multinational companies and extended enterprises. This Partnership with MCI overarches our whole international strategy.

The BT-MCI alliance (in which BT will own 75% of the proposed joint venture) is pro-competitive. It bridges the Atlantic; so it gives Europe a platform in the USA. And, by combining BT and MCI's technologies and marketing skills, it will create new services and result in a best of breed infrastructure able to compete with anyone in the global marketplace.

The new joint company will provide products and services for major customers around the world. It will be responsible for the end-to-end global service aspects of the product portfolio. Its initial services include global Virtual Network Services (both voice and data), managed data services including frame relay, global bandwidth services, global visual services and messaging.

I have no doubt that other companies will follow the example of BT and MCI to form global alliances geared to serve customers better. But the speed with which they will be able to respond to their customers needs is inextricably linked with the willingness of operators and regulators (both national and international) to open their markets to competitive forces.

Progress has been slow. But, I am sure many of you will have heard me say before that if the customer wants a service and the technology can provide it, sooner or later the regulator will allow it. I believe it is vital for Europe that it is sooner not later.

Radical liberalisation of markets in the European Union is central not only if customers to be offered competitive products and services but to ensure that EU based telecommunications companies survive into the next century. Let us hope that Europe responds, and quickly.

Chances of France Telecom in a Global Market

(This lecture was originally held in French language)

Marcel Roulet

Ladies and gentlemen,

I am delighted to be here today with you in Munich at an event which brings together the members of the telecommunications industry -- manufacturers and operators alike -- along with their customers.

By choosing to entitle this conference "Global Players in Telecommunications", the Münchner Kreis has very accurately focused our discussions on the key issues before us today: the increasingly fast-paced globalization of the economy and the far-reaching changes currently taking place in the telecommunications market. The major challenge facing world-class operators is successfully making the strategic adjustments dictated by these two developments. France Telecom is keenly aware that only the players who are able to make these adjustments will enjoy long-term success -- and we intend to figure among the "winners" in this market.

Although this may be somewhat of a simplification, I think many of my colleagues would agree with me that the structure of the market is currently moving towards three main categories of operators:

- There are a handful of major operators -- or groupings of operators -- who offer a diversified range of services around the world.
- Then there are a good many companies specialized in sometimes very profitable market segments, such as mobiles.
- Lastly there are regional operators, which have generally grown out of former national monopolies and which provide their services to the two preceding types of operators.

It is abundantly clear that France Telecom seeks to establish itself in the first category -- as a major global operator. What I would like to address here today is why and how we intend to achieve this position.

To start with I would like to give you a rapid overview of the France Telecom group, befor turning to my vision of where the telecommunications industry is headed.

I would then like to give you a rapid overview of how France Telecom is working to serve its customers, not only in France but around the world. This of course includes on of our company's most fundamental strategic moves in recent years, our partnership with Deutsche Telekom, which would also like to review for you.

1 Positioning of the France Telecom Group

1.1 Key figures

France Telecom is first and foremost a very large customer-driven enterprise. A few figures will illustrate this:

- We have revenues of about 23 billion dollars, which makes us France's eighth largest company and the world's number four telecommunications operator in terms of revenues -- third in terms of international traffic.

- Our cash flow totals seven billion dollars,

- France Telecom has an installed base of nearly 31 million main lines, 6.5 million Minitel terminals, 200,000 public phones and 600,000 ISDN B channels.

- The Group has a consolidated work force of 165,000 persons.

I think that these figures indicate our capacity for action and the weight of our group. These figures also explain why we are confident in our ability to meet the challenges we face.

1.2. Status

Let me take a moment here to discuss our status. France Telecom is currently a state-controlled company. In order to achieve the objectives it has set, France Telecom aims to become -- and indeed must become -- a full-fledged enterprise. I have initiated a broad internal debate on this issue within France Telecom, and will present our conclusions to the French government to allow it to make the appropriate decisions.

1.3. A European enterprise

France Telecom is also a resolutely European company. This is obvious from a geographic perspective, and it is becoming equally evident from an economic one. The development of telecommunications is driving internationalization, and Europe is a natural step along this path to global development. From a legal standpoint, this will become a reality as of January 1, 1998, the date on which all telecommunications services in the European Union will be opened to competition.

This will make the twelve nations of the European Union -- which will probably have grown to sixteen by then -- the domestic market for the operators of each member country. In reality, though, the opening of this market will not be quite this simple, since certain countries will be accorded additional time to adapt to competition. The process of European unification has been initiated, however, and we are delighted with this development, since this is what our customers want. Both major international firms and private customers want equipment, services and tariffs which give users access to the entire zone covered by the European Union, wherever they might be at a given time.

2 My vision of where the telecommunications industry is headed

2.1. I would now like to share with you what I believe are the conditions which must be met to allow harmonious development of the telecommunications market, not only in Europe but at a global level as well.

2.1.1. The rules of the game

I think that the first condition is that there be clear, long-term and balanced rules governing development.

An industry such as telecommunications, which is so vital to the world economy, simply cannot be subject to the law of the jungle. This industry requires rules which are coherent with underlying economic and technical realities. The last thing it needs is the sort of nit-picking legalism which may make lawyers happy, but which leaves consumers by the wayside.

These rules are established by national regulators, but for Europeans it is evident that Brussels now plays a dominant regulatory role. We feel it is our duty to play a constructive and realistic role in the definition of these rules whenever we have an opportunity to do so. We are particularly vigilant in this regard since, as everyone is aware, France applies European directives to the letter as soon as they have been approved. At the same time it is only natural that a sector such as ours, with its pivotal role in the globalization of the world economy, must also have a global structure and be governed by rules which transcend national or regional markets.

This is one of the key issues to be dealt with in the GATT negotiations. We sincerely hope that it will be possible to define healthy and fair rules based on two essential principles: reciprocity and equal treatment, excluding any one-sidedness to the detriment of Europe.

2.1.2. - Harmonization

In addition to rules governing fair play, the telecommunications needs harmonization, which is the second essential condition I would like to examine.

And I think we should be very clear here in stating that harmonization and competition are in no way incompatible. On the contrary, they are two complementary aspects which, when correctly balanced, provide the best foundations for customer satisfaction. The need for this balance is implicit in the decisions adopted by the European Community at the end of 1989, although in the past two years one has the feeling that the goal of broadening competition has relegated the need for harmonization to a secondary role. In this regard I cannot help but feel concerned when I see attempts to use the Open Network Provision as an instrument for accentuating competition. ONP was in fact conceived as a means for harmonization, by facilitating and structuring fair access for service providers to the networks owned by monopoly operators.

It would be extremely dangerous to lose sight of this harmonization objective, since the future of our networks depends on it. The quality of the services offered by networks -- whether telecommunications, transport or energy, for that matter -- depends on the quality of the interconnections and the measures taken to guarantee interoperability. This can only be achieved on the basis of reliable technical standards and healthy economic conditions. If these technical conditions are not satisfied , interoperability will not be effective and users will not be happy. And if the fees paid to the network operator are insufficient, the quality of service will decline, again leading to consumer dissatisfaction.

This is why France Telecom -- like most European operators and in particular our partner Deutsche Telekom -- has insisted on this extremely important and often poorly understood aspect of building a new context for the telecommunications industry.

2.1.3. - Cooperation

The third and last condition to harmonious market development is cooperation between operators. Here again, I think it worth emphasizing that cooperation and competition are not incompatible, even though I reject the argument that cooperation is more acceptable if the participants have been engaged in aggressive competition with one another....

Above and beyond the tradition of mutual support and indeed amicable relations which reign in the telecommunications community -- a tradition which has long been an essential component of France Telecom's corporate culture -- it is our firm conviction that it is in the interests of all concerned to pursue balanced and durable partnerships. This philosophy governs our work in Europe, where France Telecom systematically participates in major multilateral projects.

Among the most important of these are:
- GEN, the Global European Network, which is the trans-European platform for leased circuits;

- Metran, the Managed European Transmission Network, an SDH transmission platform for all applications;

- the ATM Pilot;

- Euro ISDN, comprising an Integrated Services Digital Network conform to ETSI standards;

- the European Telematic Network, for cross-border database access;

- and of course GSM digital cellular telephony, which is no doubt one of Europe's most brilliant success stories in terms of cooperation in the telecoms sector. Indeed, this success continues to expand today far beyond the borders of the "Old Continent".

2.2. *Having looked at the three conditions I feel are essential to the harmonious development of telecommunications, I would now like to consider the question of infrastructure development.*

The spirit of cooperation which we promote naturally led us to welcome the emphasis placed by Article XII of the Treaty of Maastricht on the importance of trans-European networks. This is an innovative concept which Jacques Delors subsequently developed further from an operational perspective in his White Paper. More recently, a group of economic decision leaders was invited to define a concrete program of action for the European Community for the coming years.

At the same time, US vice president Al Gore has laid out the directions to be promoted by the American government for the country's National Information Infrastructure, which is designed to enhance the competitiveness and economic leadership of the United States.

You no doubt share my annoyance at times with the sensational media reports following the host of "information highway" announcements in recent months, each trumpeting billions more dollars and more gigabits!

Happily, a number of very lucid industry leaders, such as Bill Gates, have refocused this debate on a more rational level, which is precisely what I would like to do.

Obviously, when you are dealing with complex subjects such as these, which combine advanced technologies with mass utilization, there will be a certain amount of political and media involvement. At the same time we have to avoid the temptation to follow fashionable trends, which can quickly lead to disappointment, particularly when it comes to the impact on employment. We need to be very specific in defining what we are talking about and we need to clearly define precise objectives and roles.

The notion of "network" is extremely rich, but it should not be a source of confusion. For example, things which may be true for energy or transport systems may not necessarily apply to information technology networks:

- In the first case, the emergence of international networks by interconnecting existing domestic networks is primarily the result of major cross-border investments, and the domestic components have been interconnected without too much difficulty.

- For telecommunications networks, on the other hand, interconnection does not require heavy investments. What it does require is the implementation of complex processes, standards, specifications and software, which in turn impact on the national networks themselves. In this case, the primary need is an efficient and consistent cooperative framework.

The issue of what type of regulatory environment is most propitious to the development of information networks is fundamental. Some would affirm -- keeping in mind their own personal agenda -- that speeding up the process of liberalization being pursued in Europe is itself the primary condition for ensuring the success of trans-European networks. This position, however, calls for two observations.

To begin with, as I noted a moment ago, companies which are investing in infrastructures must have clear visibility of the regulatory framework which governs their business. For Europe, the schedule defined in June 1993 calls for complete opening of all services to competition in 1998, along with a review of rules applicable to infrastructure based on a Green Paper to be submitted before the end of 1994. The in-depth and objective analyses contained in this Green Paper cover every aspect of the issue and are to serve as the basis for the global solutions to be implemented.

Secondly, we must take a very lucid look at what has happened in markets which have already been opened to competition. In the United

States, despite aggressive liberalization of the telecommunications sector for the past ten years, we still see that strong political initiatives are needed to support the implementation of networks which have not been created "spontaneously" by a deregulated market.

Likewise, in the United Kingdom there has been absolutely no cause and effect link between the development of competition and accelerated implementation of broadband networks.

On the other hand, both of these markets reveal the pernicious effects of asymmetrical regulations which, by imposing specific restrictions on the dominant operator, have hampered the growth of multimedia.

But this does not mean that telecommunications operators can simply sit back and demand their share in trans-European networks: they also have to assume the corresponding responsibilities. Europe's operators have not of course been idle and already have a number of impressive achievements to their credit, including the major international projects I mentioned earlier. Furthermore, some operators have already created national networks which can be interconnected with those of their neighbors to create very solid foundations for a European superhighway. France Telecom, for example has built the Renater high-speed research network and an ISDN infrastructure which spans the entire country.

Still, we must recognize that telecommunications operators must go even further in accepting their responsibilities if we are to establish our credibility in the eyes of the public.

Doubts continue to be expressed for a number of reasons. It still takes too long to bring new projects to fruition, leading some observers to question the real priorities of the operators.

In addition, rates are still perceived as too high -- although the causes for this are recognized -- and even dissuasive when it comes to international links.

The operators can and must rectify these shortcomings. And yet an improved offering driven by advanced technologies will only be truly effective if there is dynamic demand for these new services.

In order to structure and stimulate new uses of telecommunications resources, we must see coordinated action by governments and the providers of these services.

It is natural that governments intervene in areas related to "institutional" or public service needs, such as education, health care, research or cooperation among different government entities.

As for the service providers, their responsibility is to foster the emergence of market-driven solutions to the needs of both consumers and business customers. It must be abundantly clear that the mission of telecommunications operators includes direct participation in defining and implementing concrete projects for information transport, integrating the contributions of other players. The operators are not simply "pipeline suppliers": they also have a legitimate role as international providers of services with high added value.

3 Customer-focused actions pursued by France Telecom

3.1. France Telecom is preparing to meet the challenges of growing competition

Before I turn to the international growth of the France Telecom Group, I would like to take a brief look at some of the things we are doing to meet the ever-increasing competition we face in our domestic market.

These actions focus primarily on service quality, customer segmentation, reorganization of our commercial network and, last but not least, on restructuring our tariffs.

- A strong commitment to quality

We have for the past eight years pursued a major program known as "Quality +". This effort, which is based on surveys of our customers, has led to Group-wide mobilization in order to improve the quality of our services.

We are also preparing to meet our public service commitments in terms of service quality for 1995 and we have made organizational changes which give greater responsibility to "front line" employees who have direct contact with our customers.

- Customer segmentation

We are pursuing major initiatives which enable us to target our actions to precisely defined customer segments. Our three main segments are: major corporate accounts and large businesses; smaller businesses, including "professionals"; and residential customers. In addition, our strategic marketing and sales actions target about twenty different sub-segments.

- Reorganization of the commercial network

We have also undertaken an in-depth reorganization of our network of sales offices. This network now consists of:

* a department which manages our 300 largest corporate accounts;
* 18 business sales offices which handle the accounts of 30,000 multiple-site business customers;
* and 200 sales offices to serve our three million small business and residential customers.

- **Restructuring of tariffs**

France Telecom's new rate policy is designed to meet the challenge of greater competition and resale.

Our aim is to build loyalty among our customers and to diversify our rate structures according to different customer segments.

This restructuring has involved a major reform of tariffs in early 1994 based on larger local calling zones, coupled with continued rebalancing of local and long-distance tariffs.

We also continued to lower our international rates, which dropped an average of seven percent for telephone use. There have been even more significant rate cuts -- on the order of 14 percent -- for links with numerous European countries.

As you can see, France Telecom is pursuing a resolutely offensive commercial strategy in our domestic market. At the same time, these actions are being extended in international markets to enable us to serve the international needs of our customers.

3.2. <u>The France Telecom Group's international expansion</u>

France Telecom is making the investments needed to develop our business volume throughout Europe and the rest of the world, especially in countries with strong development potential. This commitment to proactive international development was in fact one of the primary goals stipulated in our contract with the French state for the 1991-1994 period.

As a mainstay carrier, it was and will continue to be essential for us to build on our traditional core domestic operations, in order to expand our reach to the rest of Europe, and indeed the rest of the world.

3.2.1. France Telecom around the world

With offices around the world and a constantly expanding commercial and operational network, France Telecom is now present throughout the five continents. Whether in North and South America, Europe or Asia, the missions of these locations extend beyond simply establishing our presence in foreign markets.

The experience we acquire locally significantly enhances our understanding of these markets and enables us to learn very concrete lessons from the "best practices" pursued in other countries. What's more, this active presence in the field enables us to regularly seize new market opportunities, to judiciously select promising projects and to establish operations that will prove profitable over time.

Of course we could not successfully expand into new markets without first having acquired a complex pool of in-depth expertise in our domestic market. For example, our recent success with GSM digital cellular systems has enabled us to export this technology to Greece and India.

We also gained unique experience in France in rapidly modernizing the basic network infrastructure, having tripled the size of the French telephone network between 1974 and 1981. This has enabled us to transfer this expertise to a number of countries seeking to develop their telecommunications infrastructure as a means to support and drive expansion of the overall national economy.

In Argentina we have teamed up with the Italian operator STET and the local group Perez Companc within Telecom Argentina, which operates the telephone network for northern Argentina.

In Mexico, we are part of the operator Telmex, alongside our two partners, the Mexican group Carso and the North American operator Southwestern Bell. Telmex is operating and developing Mexico's national telecommunications network.

In Russia, we have joined Deutsche Telekom and US West in signing a partnership agreement with the Russian operator Rostelekom to develop a digital network for the Russian federation serving long distance, intercity and local needs.

In Poland, we are working in conjunction with Ameritech to offer a cellular phone service which already had some 15,000 subscribers by the end of 1993.

More recently, I traveled to Vietnam to sign a memorandum of understanding with the Vietnamese telecommunications operator VNPT, covering a global long-term partnership in that country. With its position in the heart of Asia and its unique potential in this region, Vietnam can be considered as a key path to further development in this part of the world.

3.2.2. Moving towards global solutions

As we focus on our objective of providing global solutions for our customers, France Telecom enjoys a full range of assets, recognized by the great majority of telecommunications industry observers.

We are one of the few operators in the world to offer such a complete and constantly evolving range of products and services. Our offering covers all the communications needs of our customers: fixed or virtual corporate networks, leased lines, mobiles, telephony and a host of advanced ancillary services made possible by digital technology.

France Telecom's catalogue calls on state-of-the-art technologies such as fiber-optic lines and satellite links. Our Group also enjoys the support of the CNET, a world-class research center and the largest telecommunications research facility in Europe.

But these advantages would not mean as much if they were not deployed within a resolutely offensive strategy aimed at attacking world markets on several fronts. Through our Transpac subsidiary in particular, the Group is taking steps to expand its offer and provide a uniform data network covering all of Europe.

We also continue to launch new subsidiaries as we focus on meeting the international needs of our customer businesses by establishing our own solid global position. Because what these international players want above all is access to a global network that frees them from operating constraints and multiple standards.

There are a number of indicators which lead us to be optimistic about our ability to serve the international business market. One which I would like to mention here is a survey done by the Louis Harris Institute last year on the attitudes of business leaders in eight Western European countries towards the liberalization of the telecommunications sector. One of the questions was whether they were satisfied with the current level of service. We were delighted to note that, of the eight countries polled, the level of satisfaction was highest in France, which received a rating of over 80 percent. Our recognized ability to satisfy our most demand customer segment is, in my opinion, one of France Telecom's strongest advantages in our quest to become a global player in a global market.

At the same time, it is widely accepted that in the emerging telecommunications industry, no operator can alone achieve its strategic objectives. Both France Telecom and Deutsche Telekom have long subscribed to this belief.

Other players have come to similar conclusions, albeit somewhat later. Today, though, the facts speak for themselves, and one has but to consider the announcements made in 1993: BT and MCI, AT&T and McCaw, US West and Time Warner, Bell Atlantic and TCI -- even though this last proposed link-up recently fell through.

Whatever forms these *rapprochements* might take, we have clearly entered the "era of alliances". The underlying climate of competition behind these moves has provided further incentive for the offensive strategy we are pursuing in order to fully satisfy our customers.

This is why I am particularly pleased to be able to address this gathering here in Germany today alongside Helmut Ricke, at a time when the alliance between Deutsche Telekom and France Telecom is taking a decisive step forward. This new phase is an extension of the longstanding cooperation we have pursued for many years. Our partnership includes our Eucom joint venture to provide value-added services to international business customers and of course Eunetcom, which provides major corporate customers with flexible solutions, including partial or complete outsourcing of their networks.

But we intend to achieve more than simply taking another cooperative step forward: our goal is to pool the very best of our resources in order to provide the greatest possible benefits for our business customers. This will enable us to offer them the full range of telecommunications services currently open to competition, everywhere in the world. The scope of competitive services will very likely expand considerably, and our new company will be in an excellent position to progressively expand its operations, entering promising new segments of the market.

Let me very briefly give you a look at the shape of this alliance, which has of course been defined in close cooperation with Mr. Ricke. Our new company will be created from existing entities contributed by each partner, and our goal is to be operational at the beginning of 1995. We aim to achieve maximum commercial impact by merging our sales and marketing teams around the world, as well as the Transpac and Datex-P networks, at both the national and international levels.

We are committed to building a partnership which benefits Europe, by giving it a world-class operator capable of facing the global competition.

I will conclude by emphasizing that this alliance is a major component in France Telecom's international strategy. We are making a long-term commitment to its success and we recognize that we are attacking a very difficult market. And yet, like Deutsche Telekom -- or more precisely in conjunction with our partner -- we have the advantages and resources needed to succeed. Both members of this alliance have build up a relationship based on confidence and mutual respect, and we share a common vision of the future of our industry. What's more, our two organizations are very complementary.

Today, we are launching a major contender in the European and global markets. Tomorrow, we intend to be a benchmark player in these markets.

Thank you very much.

New Opportunities Through Cooperation

Helmut Ricke

- The spoken word shall prevail -

Ladies and gentlemen,

In our sector of industry, alliances, partnerships and other forms of cooperation are being forged at an unprecedented pace.

It is a pleasure for me to have an opportunity to discuss the topic "New opportunities through cooperation" with you and to present Deutsche Telekom's views on this subject.

Here, I do not want to take a strictly pragmatic approach and limit my comments to the compelling need for partnerships. Instead, I also intend to examine this subject from a different angle. In addition to this compelling need, I would like to go into two other important aspects, namely the opportunities and challenges presented by cooperation.

When two companies decide to cooperate, both companies must perceive this as a gain. This is certainly our overriding consideration when we reflect on this topic.

The alliance we formed with France Telecom was also naturally based on this philosophy of generating synergies. By pooling the strengths of the two enterprises, it is possible for us to tap new core capabilities in the fields of technology and data transmission. Most notably, services offered to key accounts and business customers can be optimized.

Cooperation does not just provide new opportunities. It also presents us, as leading enterprises, with new challenges. How our companies will be changed when we realign ourselves by forming strategic alliance networks is naturally a central issue. These are processes that need to be carefully managed.

Topics/Main Points of Speech

In the course of my talk, I will touch on three main points:

- First: How do alliances and other forms of cooperation develop into instruments for implementing corporate strategies in the telecommunications industry, and what are the driving forces?
- Second: How is Deutsche Telekom positioning itself in this rapidly changing market?
- And third: What conclusions do we draw from the experience we have gained with alliances and other forms of cooperation?

A Sector in an Alliance Frenzy

Our sector has been in an "alliance frenzy" for some time. The list of projects is so long that they cannot all be mentioned here. I will therefore just refer to a handful of typical examples: British Telecom and MCI, Viacom and Paramount, Deutsche Telekom and France Telecom, Siemens and Italtel and, finally, AT&T and Novell along with Mircrosoft.

The number of alliances and other forms of cooperation has continuously increased over the past five years - at an annual rate of just under 40 percent on average.

Wide Range of Alliances: Within a Specific Industry and Between Industries

The examples I have just mentioned already give an indication of how wide-ranging these alliances are. They are concluded not only <u>within</u> a specific branch of industry but also <u>between different</u> branches of industry.

Within a specific branch of industry usually means: alliances at the international level.

Alliances between different branches of industry that are most notably causing a stir are those with companies in the media industry.

As the multimedia develops, there is increasing cooperation between telecom carriers as well as equipment manufacturers from the telecommunications industry and the computer industry.

The major reason for forming an alliance is obvious: namely, to open up new markets, whether this occurs at the international level or by gaining access to newly emerging market segments.

Complex Networks

Above all, a trend can be seen towards complex alliance networks. Already, the web of networks of a few telecom carriers is the product of well over 100 alliances and other forms of cooperation.

Network Management

Although alliances are the order of the day in the field of telecommunications, they are not an achievement of our age. For, it was only owing to well-functioning cooperation in international bodies and organizations that it was possible to create the conditions needed to place telephone calls around the world as we can today.

What is fundamentally new, however, is the extent to which alliances and other forms of cooperation are now being established.

Given the increasing complexity of the networks that are being created through alliances, there is naturally a need for well coordinated network management to generate the intended benefits and added value. Here, I would just like to mention in passing that a finding from chaos research should serve as a warning: the more complex the system, the greater the risk that wrong decisions will be made and a wrong course taken. There is a great danger that in resolving existing problems, new problems arising from the solution will be disregarded. Then, what pyschologists call the "logic of failure" will run its course.

But this is just a critical comment on the side. I would now like to return to our main concern: the logic of success. With regard to controlling complex alliance networks, I would like to point out two problem areas in particular.

Balance: Cooperation and Competition

Many alliance partners that have entered into particularly productive cooperation in certain business segments are definite rivals in other areas. One of the major challenges arising from a company's cooperation policy is therefore the question of how to strike a balance between cooperation and competition.

It is a new and unusual relationship for carriers to cooperate with a partner who remains a competitor in other fields. This could lead to uncertainty in their dealings with each other. This is soon bound to change, however. For, companies are going through a learning process, which will lead to a new corporate concept.

This gives rise to innumerable questions. I shall mention just two examples: "How can I prevent an undesirable transfer of know-how?", or "How should I act towards a telecommunications provider with whom I am now competing, but with whom I would like to cooperate in the future?"

Together, we will have to find answers to these and other questions.

But one thing can be said for certain. To effectively control alliances, we will need clear-cut agreements and close monitoring. Only then will it be possible to avoid misunderstandings and misinterpretations and to realize in time if a wrong course is being taken.

Greater Flexibility and Transformation of Corporate Culture

Though I speak of potential problems that might arise in the course of cooperation between partners, the opportunities still far outweigh the critical aspects.

I have already pointed out that companies cooperating with each other are able to tap new core capabilities by pooling their respective strengths and that another advantage is the possibility of optimizing the services offered key accounts and business customers.

Moreover, alliances and other forms of cooperation release energies that help promote greater flexibility and transform the corporate culture. Here, I would just like to mention two aspects, as an example: breaking away from traditional corporate values that are in need of being overhauled, and creating flexible, agile and unbureaucratic units that are, so to speak, speedboats flanking the fleet of large allied companies. I feel that these two aspects are particularly important. Our alliance policy cannot be dedicated merely to the concept of sheer size. It is not a matter of creating ever larger corporate structures, but of having better, more efficient ones. And "better" today means, as a rule, faster, more flexible and better able to adapt.

In this context, I would like to just briefly mention "TeleCash". Only by pooling the strengths of the joint venture partners, IBM and Deutsche Telekom, and creating a flexible unit was it possible to successfully establish this service on the market.

Background and Driving Forces

Now that we have dealt with the opportunities and risks presented by alliances and other forms of cooperation, let me go into the question of why alliances are playing an ever greater role in the implementation of corporate strategies.

What are the driving forces behind the "feverish activity to form alliances"? And what changes are occurring on our markets?

To begin with, the telecommunications market is extremely attractive and has an enormous potential for growth. In 1993, the world market for services and hardware was worth just under 510 billion US dollars according to OMSYC figures. This is a nominal increase of 16.5 percent over 1992. Europe accounted for some 36 percent of global telecommunications services sales. For this reason, and naturally also owing to its extremely important geographic location from a strategic point of view, Europe is becoming one of the major arenas of international competition.

Structural Change

The telecommunications market is currently undergoing a dramatic structural change. In my view, there are four decisive factors here:

- Changing customer requirements,
- Increasing liberalization,
- The globalization of markets and
- The convergence of technologies and markets.

There is no need for me to go into any further details here before this audience about the challenges presented by liberalization and globalization. I intend to discuss later on, anyway, how Deutsche Telekom is preparing itself for these developments. But just let me touch briefly on the first and fourth point.

Changing Customer Requirements

Over the past decade, customer requirements have shifted strongly in the direction of integrated global networks. Large companies have increasingly been going global and, in addition, are becoming more and more integrated electronically. Instead of being oriented mainly to packet-switched data transmission within their own country, as was the case in the past, key accounts now more than ever require the facility management of complex networks.

Convergence of Technologies and Markets

There is thus a causal relation between this and the convergence of technologies and markets. We are evolving in the direction of integrated data, image and voice transmission, and our business customers are increasingly expecting to receive an integrated offer in one-stop shopping. As a result, markets - such as the computer industry, the media sector and telecommunications - are converging to form one overall market, and new market segments are emerging. The prime example here is "multimedia".

New Demands Placed on Telecom Carriers and Service Providers

These market changes are placing new demands on telecom carriers. I will mention three of them:

- First: a global market presence to do justice to the needs of all customers. When I say "all customers" here, I am not just referring to business customers, but naturally also to our residential customers.

- Second: the entry into new, complex markets going far beyond the traditional core capabilities of a telecommunications network operator and service provider. Again multimedia is a prime example of this.

- Third, investment in the development of new technologies and the accelerated installation of new digital networks.

International cooperation is becoming a conditio sine qua non for each and every one of us. In order to operate successfully on the market, I see a need for telecom carriers to promote joint actions on the part of market players from all areas in order to ensure that the core capabilities of all partners will be geared towards making use of synergies.

We at Deutsche Telekom are convinced that these new challenges can only be met by fundamentally reorienting our strategic policy. This transformation process is already in full swing and is reflected in our most recent corporate decisions.

Deutsche Telekom's New Corporate Identity

The new corporate identity with which we will be entering the next millennium can be put succinctly: The enterprise Deutsche Telekom is being transformed from a nationally oriented network operator into a globally oriented network operator and service provider. This means that we are concentrating on offering a comprehensive range of services at a higher value added level, naturally without neglecting our network operator function.

It is our aim to firmly establish ourselves as one of the leading global players. In so doing, we intend to transform our enterprise in close partnership with international telecom carriers.

We are positioning ourselves at two levels reflecting the structure of our customers; on the one hand, as a service-oriented, innovative and highly efficient provider of services for the mass market and, on the other, as a provider of integrated system solutions with modularly structured services for international business customers.

Three Strategies

We are pursuing three strategies that are the strong driving force behind the corporate transformation of Deutsche Telekom. I would like to just briefly mention these:

- First: safeguarding our core business on our home market;
- Second: opening up future growth markets,
- And third: the internationalization of our business activities.

Safeguarding our Core Business

Safeguarding our core business on our home market appears to be the natural thing to do. And yet, following the EC's liberalization decision, this presents a great challenge for all European telecom carriers, with the exception of the British, where this development is much farther advanced.

We here in Germany are therefore relying on, most notably:

- Further improving the quality of service,
- Obtaining additional traffic,
- Increasing our productivity while exhausting all possibilities of lowering costs and
- Especially, selectively upgrading our infrastructure as well as
- Creating new service offerings.

Opening up Growth Markets

Our second strategy involves opening up growth markets at an_early date, and here I would expressly like to emphasize "at an early date".

One growth market that is being widely talked about is multimedia. With regard to professional applications for business customers, this involves the integration of all forms of information, while in the residential customer area, the focus is on entertainment, that is, the audiovisual aspect. Deutsche Telekom is well-equipped for developments in both areas.

Business Multimedia

Business multimedia will evolve in the direction of multifunctional, interconnected workstation systems. Deutsche Telekom is already working closely together with the respective hardware producers in order to put terminal equipment on the market at mass market prices. That multimedia services are more than just a future dream can be seen from the videoconference service, which already has sizable growth rates.

1. Consumer Multimedia

As is the case with business multimedia, conditions for entering the consumer multimedia market are also favorable. Deutsche Telekom has Europe's largest broadband cable network with 13 million subscribers, which is even larger than that of TCI, the largest cable network operator in the USA.

We have developed a strategy for optimally positioning ourselves on this multimedia market. The focus is on the transport of services. The broadband cable network, satellite links and both ISDN and telephone networks are available as transport paths. We attach particular importance to establishing a strong position in the distribution of programs via satellite.

2. Mobile Communications

I would now like to turn to another market that is already booming today. I am speaking here of the growth market, "mobile communications".

It is clear even now that mobile and personal communications will be the form of communication of the future. Telephone customers will have personal numbers with which they can be reached anywhere in the world, regardless of where they are, either via the radio communications network or via satellite.

Owing to our dual position as an operator of both fixed networks and radio communications networks, we are well equipped to operate successfully on this promising market of the future.

Working from a position of strength, we are concentrating on quickly penetrating markets with strong commercial partners to assist us in marketing and distributing cards and terminal equipment.

Furthermore, we can make use of our know-how by entering the market as a service provider. This means that we will take over tasks in the areas of "marketing", "sales" and "administration".

Internationalization

With regard to our third strategy, internationalization, we differentiate between two things. First, gaining access to international markets that are relevant for key accounts and business customers, namely, in the triad of North America, Eastern Asia and Europe. Second, engaging in regional diversification to take advantage of market opportunities currently presenting themselves, most notably, in Eastern Europe. Here we see ourselves playing the role of a network operator as well as that of a service provider.

I would like to limit my comments here to the first aspect, access to international markets for key accounts and business customers. We will take into account the individual needs of this group of customers, which require cross-border voice, data and image communication, by offering them completely integrated solutions in the future. Our motto here is "one-stop shopping" and "one-stop billing".

International Network Strategy

The prerequisite for this is an internationally oriented network strategy. Here we are focusing on setting up and operating:

1. Data networks and global virtual private networks,
2. GSM-based mobile communications networks, including participating in international consortia and
3. Fixed networks in the form of participating interests and joint ventures, making global network operations possible.

This international network strategy is naturally an important basis for entering into new alliances and other forms of cooperation.

Alliances as an Instrument for Implementing Strategies

Following this overview of Deutsche Telekom's strategies, I would now like to give you an idea of how we intend to implement them and of the importance that "alliances" will have for us in this process.

One very important instrument that we are using to implement our strategy is to form strategic alliance networks providing blanket coverage. The opportunities that will emerge as a result of such networks are obvious.

By forming alliances we can ensure that our customers will have access to integrated services being offered in the entire triad. Only through alliances will we be able to develop new markets - such as multimedia - and upgrade networks worldwide. For, the need for know-how and investments is so great, that no single player can handle this on its own, not even any of the "major players" in our industry.

It also makes sense to participate in consortia to take advantage of new market opportunities opening up in Eastern Europe in order to share the investment risk that is, by nature, greater in these countries.

Following this discourse on our strategies, ladies and gentlemen, I would now like to briefly take stock of what Deutsche Telekom has so far achieved in this field.

The Alliance with France Telecom

The far-reaching strategic alliance we formed with France Telecom late last year is naturally at the center of our cooperation policy, an alliance with which we intend to pool our strengths. Forming this partnership seemed to be the logical thing to do and is the consistent continuation of our already traditional bilateral cooperation.

The European Idea is the basis for this alliance. A global provider is being established that is neither German nor French, but rather is entering the world telecommunications market as a European enterprise. The two parent companies will contribute only those operations to the Brussels-based joint enterprise that are already open to competition in their own home markets. The services offered will range from packet-switched data services, frame relay, international leased lines and virtual private networks to satellite-based VSAT services. In addition, back-bone networks with a uniform network management system are also offered, which also includes the development of a joint platform for intelligent networks.

The joint venture's relevant share of the global market will be under 10 percent. Despite the substantial investments made, this enterprise will be far from being in a position to dominate the market.

It is an open secret that we would like to expand our alliance with other partners, for example, from the USA or the Far East. There are still a few hurdles to be overcome, however. The verticle integration of a few potential partners undoubtably poses a problem. Moreover, some conceivable constellations might give rise to regulatory reservations.

Pan-European Cooperation in Other Areas

In addition to the far-reaching alliance with France Telecom, pan-European cooperation is important in a number of other instances within the scope of our internationalization strategy.

Here are just a few examples: A broadband infrastructure that is as versatile as possible is a basic prerequisite for offering cross-border, integrated services. We are counting on the ATM technology. We have reached an agreement with the other major European network operators, BT, France Telecom, Telefónica and STET, to set up a Europe-wide digital high-speed network, the Global European Network, as an important preliminary stage to a future ATM network. At the next stage of development, this network will then be replaced by METRAN, the Managed European Transmission Network.

In the United States of America, the initiative taken by the government to create an information superhighway has quickened the pace of change. We could use this spirit of change here in Europe, too, right now. I am convinced that a comparable step on our continent would have a similarly positive effect on growth, employment and the competitiveness of European industry. I therefore welcome the European Commission's idea of creating and using a high-performance Europe-wide information infrastructure.

We have already made quite some progress in this direction. European network operators and their suppliers have now moved to set up a trans-European network. Together, we have scored successes in developing standards: Here I would just like to mention GSM and EURO-ISDN.

We will be able to act on, and implement the views of the European Commission on the basis of large-scale European cooperation. This is a challenge we are all facing in the near future, which we will meet with pleasure.

Cooperation in Central and Eastern Europe

As you know, Deutsche Telekom is also cooperating in Central and Eastern Europe. What could be more natural than to export to these countries the expertise we have acquired in quickly establishing an infrastructure in eastern Germany. In our view, this involves a well-balanced strategy of support and market development.

Allow me, ladies and gentlemen, to mention just a few of the many projects that we are participating in in this area.
Take, for example, the 30 percent stake that we, together with our partner Ameritch, have acquired in the state-owned Hungarian telecommunications company MATAV. Owing to its geographic location in the Southeast Asian region, this project is of central importance for Deutsche Telekom as a strategic hub in its efforts to position itself globally. MATAV will develop into a strategic telephone traffic hub in Eastern and Southeastern Europe.

The first stage of the digital mobile communications network that we, together with our Russian partners, are setting up in the greater Moscow area is to go into service this year.

In Ukraine, we are participating in the Utel project, which, inter alia, involves setting up and operating the telephone exchange for international and national long-distance traffic.

In the same country, we have set up a mobile communications network together with our partners in the UMC (Ukrainian Mobile Communication) joint venture. I could give you many other examples.

The Goal: A Global Alliance Network

We are well aware of the fact that we still have quite some way to go before we will be able to achieve the goal of having global, strategic alliance networks. But these examples should make it clear that Deutsche Telekom has made good strides on the path to cooperation and alliances. We are open to new partnerships, but of course there must be the right "chemistry". Our partner companies must be suited to Deutsche Telekom as far as strategy and corporate culture is concerned.

Prerequisite for a Successful Partnership

In order for such a partnership to be successful, a few important conditions must be met with regard to the relationship between the two partners. It must be the aim of the two alliance partners for the joint business activity to grow instead of just focusing on one's own individual share. Furthermore, the alliance must be led by the top people in the respective companies. This is not a task for someone who has been "kicked upstairs". There must be sufficient communication, however, at all levels of the alliance. And so that it can develop, the joint venture must be allowed to have its own corporate structure.

Necessity, Chances and Challenges

Alliances, ladies and gentlemen, are neither miracle weapons nor cure-alls. Alliances are an absolute necessity in our ever more complex sector. Yet, by acting on this necessity, we open up a wealth of new opportunities in rapidly developing markets. How individual companies are able to maintain their own identity or are compelled to redefine themselves in this web of alliances will prove to be a challenge. Last, but not least, we will have to learn to master and control this new complexity.

I view alliances and other forms of cooperation as the organizational form of the future without any ifs and buts. For, companies will only be able to meet the challenges facing them if they strike out in new directions to form alliances and if they find and cultivate new forms of interacting.

This goes for our sector certainly more so than for others, for we are necessarily focusing our activities on a market comprising the entire globe.

This is an enormous challenge, but it is also a fantastic opportunity. Let's take advantage of it!

Overcoming National Boundaries

Edgar L. Brown, Jr.

Introduction

Thank you. It is a pleasure to be here today to talk about Bell Atlantic's view of the global telecommunications market, the vast opportunities we believe exist, and our strategy for participating in these opportunities.

It goes without saying that over the past 10 years, our industry has witnessed dramatic change. We are all well aware of the tide of privatization, liberalization, and competition that has swept not only the European telecommunications market, but the world.

Today we are facing the biggest opportunity that the communications industry has had since the telephone was invented.

You know there is a convergence of technologies taking place: the telephone, the television, and the computer are becoming one. Before this revolution in technology advancement is complete, your computer will speak, your TV will listen, and your telephone will show you pictures.

Bell Atlantic's mission is to be the world's leader in providing communications, information and entertainment services in attractive markets across the globe.

And we are not along in this endeavor. Major telecommunications providers and suppliers are forming strategic alliances with one another and are partnering with computing and entertainment organizations to form global alliances. Competitors are becoming partners in the drive to build what is now being called the information superhighway.

Building alliances

New alliances are being formed every day: Viacom and Paramount, Bill Gates' and Craig McCaw's Teledesic, AT&T and McCaw, US West and Time Warner, BT and MCI.

The vision is of global networks which will deliver to the customer higher value and choice in services in terms of interactivity, mobility, and connectivity, and what we at Bell Atlantic feel strongly about, flexibility.

Today I'd like to take a few minutes to share with you my view of the building of the superhighway--the road that will be our future in the information age; how others look at building that highway; and how we are looking at the highway in markets outside the United States. Critical to the building of the highway, is the building of partnerships and alliances. I want to look at some of them--and then end with my view of partnerships and alliances.

The convergence of the media and telecom industries is highlighted by the US West investment in Time Warner. With this alliance both parties win. US West gains access to a vast library of entertainment programming, and Time Warner receives a head start in its goal of having access to a fiber highway which can be used to provide homes with video-on-demand.

The proposed AT&T acquisition of McCaw gives AT&T something close to a Pan-American cellular network. AT&T views cellular as a way of achieving the growth levels it requires and of building full service networks. Wireless communication is an area of high growth in the immediate future and provides AT&T with an important link for its world-wide network.

As of just a week or so ago, the legal system in the United States has put the brakes on this deal, but the door is open to appeals.

Bill Gates and Craig McCaw, two of the United States' best-known entrepreneurs, are joining forces to promote a $9 billion communications satellite system that aims to bring the superhighway to all corners of the globe. Who knows when and if the many technological and regulatory obstacles can be overcome. However, if it can achieve this, Teledesic will be in a great position to be able to bring the benefits of the information superhighway to everybody, even those in countries with low levels of communications infrastructure.

With the United States domestic markets growing slowly while the global long-distance market is growing at an estimated 15-17 percent, which market would you like to be investing in?

BT appears to have decided that its joint venture with MCI gives the new entity a much stronger chance of winning in the global market. As we know ourselves, deregulation in your home market makes it crucial to spread your wings beyond your own shores. BT has structured the deal so that another partner can be taken on board. With a Far-East partner, BT would have strong global coverage.

The hotly fought over Paramount deal ended with Viacom the winner. With the purchase of Paramount and its warehouse of entertainment assets, its production facilities and talents, Viacom ensures that it will have plenty of high quality programming for its future cable and multimedia ventures. This acquisition is all about synergy: combining elements to add value to the whole.

In all these cases, the companies involved in alliances and acquisitions are looking for the building blocks to complete their version of the information superhighway--or to build that one section in which they are interested.

I don't know if anyone really has the "right" definition of this superhighway--but some believe it to be a whole system of interconnected networks--telephone, cable, wireless, satellite. All are upgraded to permit people to access any type of information they need or want at very high speeds.

Some visionaries refer to this capability not as a superhighway, but as a super market--one stop, one shop.

And, we see the highway paved with a full service network--that network providing a platform of wireless or wireline technologies capable of delivering any service--voice, data, entertainment or interactive--to any customer, anywhere they want it, at any time. And, it must be easy and convenient to the user--a flexible system.

Today, as the CEO of Bell Atlantic's international company, I see the new information age--the age of the superhighway--filled with opportunities unheard of in the archives of our company.

This provides us with not only new opportunities in our home market, but access to new markets across the world. The highway is not limited by borders or by oceans -- it is a technology that is available here in Europe and around the world. In fact, we are already talking to several European telephone companies about our information services and IMTV.

Leveraging success

Today, at home in Bell Atlantic's region, we are the leaders in deploying new technologies like advanced intelligent network, installing miles and miles of fiber to make our network more effective, bringing fiber to 1.25 million homes in our region by 1995 and 8 million homes by 2000, offering new services like video-on-demand, personal communications systems based on our intelligent network, and putting our entire focus on customers and their requirements as we grow our business.

We participate in the global market by building on our domestic success in technology deployment and new services to build those highways using the same flexibility and options we offer in the United States including broadband, wireless and information services.

What we consider our key to success worldwide is what we excel in here in our home markets. Stated another way, what we do well domestically, we will leverage internationally.

Now, I'd like to more fully define the Full Service Network, and then provide two examples of how we are looking outside our borders to build them.

There are three pieces, or components, to the full service network:

> First, a full service network is a distribution channel. It distributes information across the network, over fiber, over wireless, over copper wire, or over cable.
>
> Next, it is an enhanced system, combining computer systems and databases.
>
> Third, the network acts to organize and program information. Today, we watch television, for example, in a linear fashion. TV is organized by time of day and by channel. With the FSN, tomorrow, you will retrieve information you care about and watch it when you want on one channel. No video machine will be needed, no 500 channels -- just one channel with the full service network in place.

We will all see a much greater source of revenues from FSNs than stand-alone networks. And, because they will deliver higher value to customers and afford greater opportunity for market differentiation, full service network providers will have the potential to establish the long-term, sustainable competitive advantages necessary for true market leadership.

The differentiating factor in deployment of FSNs will be how we add intelligence to our network, since it is the intelligence that enables us to build a sophisticated full service network. The deployment of fiber, in which Bell Atlantic is the leader in our country, and the deployment of our Advanced Intelligent Network will support our domestic success and our global success.

For those markets outside the United States, our strategy is to build relationships, to partner with a strong company and mutually grow our businesses.

Our global partnerships

As I said, building long-lasting relationships -- partnerships -- is the cornerstone of our global success. An example of how a relationship grows and becomes a partnership is our business in Italy.

In 1989, we began a project with STET to install a large operating system to help them expand their service. Today, our European headquarters is in

Rome, the head of our European company is Italian, and we are in partnership with STET in a company in the region of Trento called Sodalia.

This joint venture company was formed to build large software systems to help each of us better manage our own companies.

Over the past years, we have worked with various companies in the STET group to develop new business opportunities and to look for ways to share our common vision. In late 1993 we announced that we will form a joint venture company with STET to offer IMTV services. Italy is unique in that there is no existing cable TV available.

Another example of deploying our strategy in foreign markets is the participation in cellular or wireless licenses and looking to build full service networks with the latest technologies providing a wide range of services to a mass market.

Most of you are familiar with our participation in the Czech Republic and Slovakia with Eurotel, our joint venture company providing cellular and packet data services in those countries.

Cellular was a way to provide a quick solution for their communications needs. We now provide cellular service to some 17,000 customers, and we are continuing to expand.

Another very important cellular acquisition took place last year when we completed our purchase of the first 23 percent of Grupo Iusacell, Mexico's largest non-wireline cellular provider, with the plan to ultimately acquire 42 percent of the group for up to $1.04 billion.

With NAFTA, economists are forecasting a 6 percent growth rate in Mexico through the year 2000. The country has only 8 lines per 100 population penetration, which means there is opportunity for real growth and the current demand is not being met by the local telephone company.

Iusacell controls about 62 million POPs including 23 million in Mexico City. And, a part of the value of this important company, is their nationwide 450 mghz license.

This license puts the company in a position to provide local telephone service using fixed wireless technologies. When the country is opened to long distance competition in August, 1996, Iusacell should also qualify as a long distance provider. We see this as a very important opportunity in a growing market.

That's a little of how we at Bell Atlantic look at the world--in terms of building superhighways in this information age. I know we are years off from a full blown information superhighway. But we are all looking at the pieces it takes to be successful. As I said when I began, putting together those pieces is critical.

Closing

Let me close by adding just a few more comments about how the pieces of the information superhighway are being put together through partnerships. No one company has the entire package to build the Full Service Network, but creating partnerships where each brings a piece and asset value to the table is a way to participate in this new opportunity.

However, let there be no doubt -- partnerships are based on the economic interests of both partners. Each has to understand the value of the relationship to the other, and the value may be broader than the one project upon which the partnership is begun.

We have learned a lot over the past eight years of being out in the world.

We've made some mistakes and we've had some successes. We've learned from both.

What is critical--there has to be a win for each partner -- and that win is in the economic benefit --that is the heart of a relationship. There needs to be a common vision, a shared view of the opportunity and mutual respect, but the bottom line--there must be value for both parties.

Partners commit and combine resources to better serve and meet customer needs and requirements.

The relationship takes advantage of new geographic and related business opportunities.

Partnerships work to create new opportunities in familiar markets and in emerging markets.

At some point, partnerships may evolve into more significant strategic alliances. That's really where, as we say, the rubber meets the road. That's where we see the key to global success. It's where there is an obligation or commitment to continually explore opportunities to work together worldwide.

A true strategic alliance enables the partners to meet mutual as well as individual objectives. You don't turn your back on a strategic partner to take advantage of a little better deal. Partners share risks and rewards. There is no such thing as an alliance where one partner benefits and the other does not.

The only limit to the opportunity will be a lack of vision and the failure to take the bold step that is needed. We believe we are equipped to do both.

Thank you.

Strong by the Joint Venture: The Unisource Spirit

Viesturs Vucins

Driving Forces

The European telecommunications industry is in a state of rapid development. The reasons for this are well documented and need not be restated in any detail, but they include the following driving forces:

- regulatory developments
- increased customer sophistication and demands
- increased internationalisation of business
- technological innovation

Driving Forces continued

- convergence of computing and communications
- economic recession
- receptiveness to technology development
- privatisation of state owned telecommunications monopolies
- competitor consolidation through alliance and acquisition

Implication of these trends

The implication of all of these trends is that the European market has entered an unsettled period. Competition is increasing, product definitions are changing rapidly and customers are continuing to demand more for less.

The total European market for telecommunications services is very promising and under exploited due to legal restrictions. It is expected to grow from Dfl 240 billion in 1992 to Dfl 430 billion by 1998

With 1998 targeted as a key year, a pan-European market is already becoming established. It will, however, take considerable time before this market becomes homogeneous and integrated, but the effects of increased competition are nevertheless already being felt by the PTOs.

Response to trends

One of the competitive responses to these pressures has been the formation of Unisource. The company was established to pool resources and hence achieve cost reductions for its shareholding companies, PTT Telecom Netherlands, Sweden's Telia, and Swiss Telecom PTT. It was also established to generate incremental revenue by capturing new market share within the emerging pan-European market.

Whilst liberalisation within the European telecommunications market is gathering pace, the only services currently open to full competition in almost every country in Europe are those related to data networking and this is therefore where Unisource and most of our key competitors are concentrating their efforts today.

By offering geographical coverage, "end to end" responsibility for seamless communication and a range of quality services that the corporate customer can not get from the local PTOs, the global service provider can win strategically important business.

Therefore the global service providers, and their shareholders invest heavily in expanding their product offerings and their global reach. They must be intent on protecting and developing the existing services for future positioning than showing short term profits.

Through its three shareholding companies, Unisource has a significant European reach from which it has built up a position as a pan-European long distance service provider, but the investment in creating a new pan-European telecommunications company with global reach is sizeable.

Global & Regional strength

Thus, a key element of Unisource's strategy is to develop partnerships. Unisource is actively seeking partnerships both on a global and regional level, and expects to finalise the key ones during this year.

Our recent agreement with SITA, operator of the world's largest network, and provider of services to the world's airlines, extends Unisource's global reach for services to over 220 countries, on one seamless network supported by one global organisation. This is very different from the traditional, bilateral way of positioning services. The SITA partnership is both a services and a commercial marketing agreement.

As announced at CeBit in March, another of Unisource's cooperative partners, Spain's Telefónica, has also decided to enter into an agreement with SITA.

Our Company- an introduction to Unisource

Unisource is a European focused company that offers a broad spectrum of international telecommunications services. It was created in order to satisfy the international demands of the founder's national customers and is a business innovator, breaking away from the traditional approaches of the single country European PTOs, and addressing the European telecommunications market in fresh new ways.

By combining the existing traffic flows of the Unisource shareholders, leveraging on the geographical "triangle" in Europe, and by optimising the networks short term, cost benefits will be derived for the shareholder and for Unisource.

As mentioned, Unisource is continuing to build its European organisation with supporting networks for business customers and enter into strategic partnerships for global reach.

Contributions to Unisource's shareholders

There is, of course, a benefit to our shareholding companies in this strategy. Two of the most important contributions Unisource makes to its shareholders are the ability to provide access to new markets and to earn new revenues in markets outside their home countries, thus offsetting the reducing market share at home.

Cost Reductions

Unisource, and any alliance, joint venture, or merger in the telecommunications industry, should provide the means and show the way to achieve savings by co-operation among the partners, or in the case of Unisource, its shareholding companies. These economies should be found in:

- shared international infrastructure
- single international market outlet for a broad range of products
- shared product / service development

- shared technology and research efforts
- shared backup capabilities
- shared administrative costs

There is a strong belief among industry analysts and major global customers that there will be only a few strong survivors in Europe which are capable of serving the global market. By joining forces in Unisource, Swiss PTT, PTT Telecom and Telia are now included among those companies which it is believed will emerge as the leaders in Europe.

This emergence as a pan-European player, coincides with the trend towards more competitive, highly customer focused industry.

Customer Requirements

Individual customers have very specific requirements and very recent surveys by Ovum, System Dynamics and IDC show that the issues at the forefront of customers' minds are: -

- cost reduction (including integration of data and voice)
- maintaining control
- improvement of efficiency and processes
- internationalisation
- breadth of services (improvement of existing services plus new services)
- seamless services

- flexibility (ease of growth, technical interfaces, contract terms, billing etc.)
- speed of implementation and service, and paying a price if you don't
- stability and strategy of supplier

In order to satisfy the European corporate customer's needs for global seamless services, the business can only be driven from a pan-European perspective. The old cliché "think global but act local" is particularly important for Unisource and our partners.

As any company builds its international telecommunications business, customer concerns must be satisfied in order to win business. Furthermore, different buying patterns are starting to emerge with multiple organisations combining to form purchasing consortia.

Success through cooperation

For example, the European VPN Users Association (EVUA) - a group of approximately 30 multinationals took the unprecedented step toward negotiating jointly for voice services. In return, Unisource and AT&T joined together in an equally unprecedented manner for European telecoms, and were jointly named an EVUA supplier for a one year trial.

Too often, companies fail to achieve their objectives not because they are unable to develop a valid strategy but because they are incapable of executing it in the single minded manner necessary. Unisource and AT&T were able to develop a cooperative strategy and secure the EVUA contract.

But Unisource is well experienced in cooperation and melding several diverse entities together for specific aims, after all this is how we came into existence.

Mission

Unisource's mission is to become a leading supplier of telecommunications services in Europe. The intent behind the mission of Unisource is to deploy the collective resources of Unisource and its shareholders in such a manner that Unisource becomes an acknowledged leader in European telecommunications in each market and customer segment it chooses to address.

The success of Unisource is dependent on a combination of early market success, global partnerships and strong support from the shareholders both financially and by transfer of skills.

Unisource is well positioned to achieve its mission to become a leading telecommunications services provider in Europe and satisfy the objectives of the shareholders.

Strong by Joint Venture - The Unisource Spirit

Viesturs Vucins
president
Unisource Group

Slide 1

Driving forces in today's European telecommunications industry

- Regulatory developments
- Increased customer sophistication and demands
- Increased internationalisation of business
- Technological innovation

Continued ...

Slide 2

Driving forces (cont'd)

- Convergence of computing and communications
- Economic recession
- Receptiveness to technology development
- Privatisation of state-owned telecommunications monopolies
- Competitor consolidation through alliance and acquisition

Slide 3

Implications of these trends

- Increased competition
- Rapidly changing product definitions
- Customers expect more for less

Net result: Changing roles and changing players

Slide 4

Response to trends

- International cooperation tending towards fewer and larger competitors
- Providing "end-to-end" seamless communications
  - One stop shopping

Unisource

Slide 5

Unisource strategy

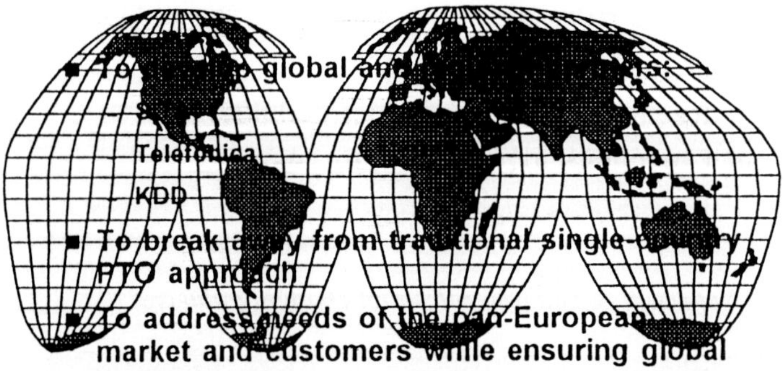

- To [illegible] global and [illegible]:
  - S[illegible]
  - Telefonica
  - KDD
- To break away from traditional single-country PTO approach
- To address needs of the pan-European market and customers while ensuring global coverage

Unisource

Slide 6

Unisource's origins

- Unisource has three shareholding companies:

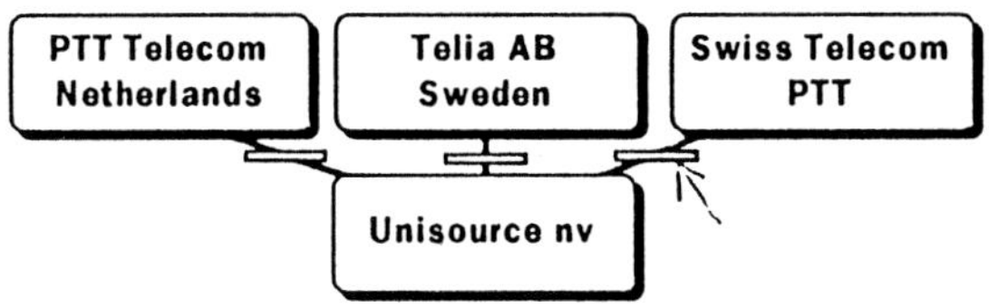

Unisource

Slide 7

Contributions to Unisource shareholders

- Access to new markets
- Access to new sources of revenue
- Cost reductions through cooperation
- Better service to existing customers

Unisource

Slide 8

Cost reductions

Cost reductions will come from:

- Shared international infrastructure
- Single international market outlet for a broad range of products
- Shared product/service development
- Shared technology and research efforts
- Shared backup capabilities
- Shared administrative costs
- Better buying power

Slide 9

Customer requirements

- Cost reduction (including integration of data and voice)
- Maintaining control
- Improvement of efficiency and processes
- Internationalisation
- Breadth of services
- Seamless services
- Flexibility
- Speed of implementation and service
- Stability and strategy of supplier

Slide 10

Success through cooperation: an example

A joint Unisource and AT&T bid recently resulted in our being named a supplier of services by the European VPN Users Association (EVUA)

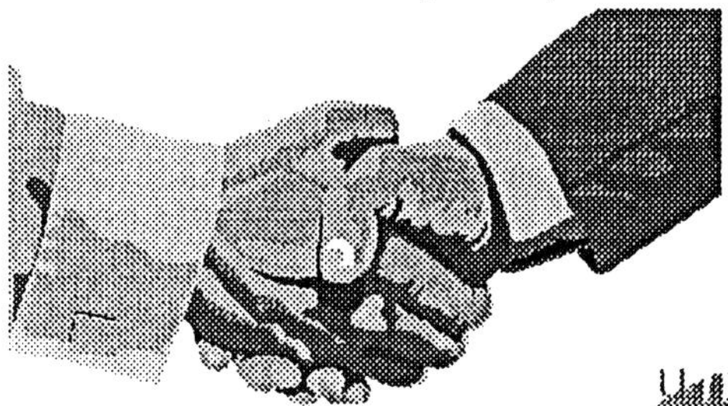

Unisource

Slide 11

Unisource's mission

- To become a leading supplier of telecommunications services in Europe

What it means to be shareholders:

- Extended markets & increased revenue
- Consolidation of resource and a more competitive stance

Unisource

Slide 12

Strategies of Leading Companies

Summary Record of the Panel Discussion Chaired by Prof. Witte

Theodor Irmer

In his introductory remarks ***Mr. Witte*** welcomed the panelists as well as the audience; he pointed out that, although discussions would start off between the panelists, it should be understood that this should by no means be a discussion only between panelists but that the audience is also invited to join in by asking questions and making comments - in short, to contribute to a discussion which should be as lively as possible. After having listened to the excellent presentations of this first day of the Congress, there would certainly be enough items to discuss or comment on. There would be no need to introduce the panelists as they had already been the speakers in today's programme.

Mr. Witte kicked off the discussion by putting forward a seemingly simple question: "What are the reasons for "going global", bearing in mind that almost all major telecommunication companies have been involved in international activities since a long time?" Why, then, all this talk on global strategies, global partners - what are the driving forces behind such buzzwords?

So far, only ***Mr. Brown*** spelled out clearly why his company extended its activities to New Zealand. It was a business decision based on economic grounds and it was not only a benefit to the company, but also the customers in New Zealand enjoy substantial reductions in their monthly telephone bills. Another example along these lines might be Telefónica of Spain, sharing in several foreign telephone companies, but when invited to this Congress it pointed out that they would not consider themselves to be a global player. For them, purchasing another telephone company would be regarded as a national matter.

So what - is "going global" perhaps merely boiling down to be a sound financial investment decision to be left to treasurers? Are there perhaps not other, additional arguments? For example, growth producing economies of scale, better use of R & D resources, more of the famous "synergies" ? Again, what is behind the trend towards globalization, what are the criteria for becoming a "global player" ?

Referring to the situation in Switzerland, ***Mr. Rosenberg*** briefly made a couple of points: Firstly, Switzerland is an international country in the true sense of the word. Its industry has a lot of international relations, the customers of the Swiss PTT are operating worldwide and, logically, the PTT has to be present where its customers are and that is, to a large extent, outside the national borders. Secondly, like other telecommunication companies, the Swiss PTT strongly depends on its international traffic which generates a large portion of its profit. Consequently, this international business has to be cultivated and maintained, particularly by drawing benefits from synergies. And, finally, in order to remain in the international game, one needs partnerships in which the partners are complementing each other in the most profitable way - staying alone will not help.

But there might also be other objectives in addition to securing market shares or making profits when going abroad; for example, investing in Eastern Europe or in developing countries is also a contribution towards supporting harmoneous development in those countries which are at present unable to help themselves but which would very well fit into a concept of business considerations.

Mr. Ricke agreed with ***Mr. Rosenberg*** that, providing telecommunication to business customers is, indeed, the driving force in globalization. As trade and business is increasingly becoming international, also business customers are becoming international customers who need not only more telecommunication services (apart from telephony), but an even more diversified range of services tailored to their specific needs. Rather than purchasing such services from, say, ten various telecommunication companies, they would like to get them through "one-stop-shopping" from one company only - and this company will therefore keep and secure its own market potential. However, maintaining this market potential is not enough, it has to be accompanied by investments in other regions, in other regional telecommunication enterprises in order to create "spheres of influence" in these regions for the future.

This was indeed the strategy of Deutsche Bundespost Telekom in the East European countries; while certainly providing support, helping these countries to overcome their telecommunication backlog, etc., Telekom simultaneously stabilized its position on the entire European marketplace, and this for a long-term period. In future, traditional telecommunication traffic routes will be changed due to the new high-speed optical fibre cables which are being installed in Eastern Europe and elsewhere; Telekom, like other carriers, is aware of the strategic importance of the re-orientation of traffic and hence, to be present in the regions in which this re-orientation is to occur.

In this context, ***Mr. Brown*** noted that his company would, above all, not invest abroad if it would not have an operating role; only if they would be allowed to participate in the management of a foreign telephone company they would make an investment. Secondly, Bell Atlantic wants to go into growth areas: in the United States - where telephone penetration is already very high - such growth will come only from value-added services, not from the traditional services. However, in markets outside the USA this might well be different as markets are emerging for either new services (e.g. video entertainment) or provide a growth potential for the traditional telephone service, mainly for residential customers.

Mr. Tarjanne added a new facette to the discussion by referring to the changing role of governments and of states in general in future. He anticipates that the existing nation state will change and that it will have to say much less than today in many areas, one of which would be telecommunication. This will also have a strong impact on intergovernmental organizations such as ITU; multinational, private companies will be playing an increasingly stronger role in telecommunication as they did up to now. He warned that, in about 10 years, the role of a state will be much less significant than it is today, a fact that should be borne in mind when talking about long-term telecommunication.

Having somewhat exhausted the discussion on "Why going global", ***Mr. Witte*** proposed to reverse the direction of this question, i.e. to analyze now who would likely go into the German market place or another country and act as a global player and, more specifically, under which rules and conditions this may take place.

Mr. Ricke courageously made a firm point by stating that competition will definitely increase when borders are open and transparent, but that a certain degree of regulation would be indispensable if one does not want to envisage the principle that everybody is permitted to do anything. Quoting the earlier example of the "open-skies" deregulation of the US airline industry which left its scars on many airlines, one might envisage a trauma that a similar situation may prevail in the telecommunication world in, say, the year 2005. Therefore, a reasonable degree of regulation would be needed; while global regulation seems to be out of reach, there should be at least an overall EU-wide regulation, complemented by national regulations in each EU-country. It is hoped that such arrangement would lead to fair and open competition. ***Mr.***

Ricke stressed that Telekom would be ready to such open and fair competition, contrary to the common belief that is still more in favour of its monopoly role which is, in fact, more an increasing burden than a benefit; hence, his clear support of competition under fair rules and conditions.

Supporting this position, ***Mr. Rosenberg*** mentioned that, although adhering to the principle of competition, he nevertheless feels that the world "monopoly" has almost become a term of abuse. A monopoly must not necessarily be inefficient and "customer-unfriendly", and the Swiss PTT is displaying low tariffs and excellent services in some areas where it still has a monopoly for the time being. But in his opinion, competition will be the better choice; it makes a telecomunication company fit, it opens the mindes of their staff but, like ***Mr. Ricke***, he stressed that competition must be fair and open, or, in his words, "a battle with equal arms".

Mr. Roulet underlined that, although at present telecommunication companies may well acquire market shares abroad by becoming engaged in specialized activities like e.g. virtual private networks, corporate networks, etc. they may do this without leaving their national territory. However, the question is what the situation would be like in Europe after 1998 when the voice monopoly will come to an end, when different national and European long-distance carriers will start competing with each other, each having their own infrastructure. In such a situation, a telecommunication company must compensate abroad the losses it is suffering from in its national territory.

Turning the discussion again to global aspects, ***Mr. Witte*** raised the question on "What would be the impact on developing countries through this new situation?". Most of the telecommunication companies going abroad concentrate their investments in growth regions; their strategies and models are tailored towards envisaged competition between themselves and established companies in these regions. Obviously, a new global telecommunication order must be developed. What would be the role of ITU, being a global telecommunication organization uniting both, developed and developing countries? Could ITU help to establish this new order?

In his reply, ***Mr. Tarjanne*** pointed at ITU's limited resources available for such a big task. However, apart from the scarcity of resources, some people even challenge ITU's role; there are opinions that the new World Trade Organization (WTO) or OECD might be better equipped to take up this issue. He finalized his statement by making a plea in favour of the developing countries; altough the borderline between developed and develping counries is somewhat blurred, ***Mr. Tarjanne*** encouraged investments by global players in underdeveloped areas in the interest of growth and development and, indirectly, economic and political stability in these regions.

Taking up this latter point, ***Mr. Witte*** noted that, although a new world order seems not to come up soon, there are at least some similarities between developed and developing countries: for both, interconnection of business centres is high on the agenda for global players, and therefore there may well be business opportunities in underdeveloped areas, although at a limited scale.

Mr. Vuccins pointed out that the trend towards competition is also very much supported by customers as they are asking for choice, better quality of service and better prices, something that was somewhat neglected by the traditional telecommunication operators. If a newly established company meets these requirements, it can well pick up a lot of business while the market share of the traditional carrier will decline.

Mr. Witte remarked that the discussion on future regulation of telecommunication was still open. With a view to the Treaty of Rome he has some doubt as to the legal competence of the EU for such regulation. However, ***Mr. Ricke*** stressed that such regulations are, in fact, already at least partially on their way; the liberalization of voice communications in 1988 is, indeed, already an early regulatory EU decision.

Mr. Claus addressed his question to ***Mr. Brown***, wanting to hear from him about experiences with deregulation in the USA which is mainly marked by competition of long-distance carriers (e.g. AT&T, MCI, Sprint, etc.) which are not permitted to provide local services. On the contrary, the "Baby Bells" (like Bell Atlantic) have the monopoly for local services but they are, on the other hand, barred from providing long-distance services.

In his reply, ***Mr. Brown*** advocated opening of the market, local as well as long-distance, for everyone, hence not favouring the current artificial structure of the industry. This would have to be supported by cost-base pricing and it would necessarily change the existing settlement scheme for long-distance services. The important elements are those of competition and free market entry, so that the market place may decide on who are the winners or loosers. In closing, ***Mr. Brown*** indicated that his model would look like this, without specifically indicating of what would be best for Europe.

Mr. Roulet stated that ***Mr. Brown's*** idea would be very interesting an analysis of the evolution in the Unites States and in Europe. It is evident that Mr. Brown's philosophy would lead to different competing networks which are supplying the services to the customer terminals. He finds such consequences somewhat paradox at a time where single optical fibres will permit to carry multiplicities of megabits directly to the customer's premises. But even if there were such an opital fibre network, there would be problems if the services would be provided by several providers. Here should be mentioned only the managment of such a "multi-service" network as one of the problems, and there are many others: open network provision, open network architecture, all of which require detailed studies.

Mr. Witte took up ***Mr. Roulet's*** intervention and agreed that the problem would be less serious when considering the interconnection of business centres. For this purpose, digital high-speed networks could be set up quite easily - in contrast to the connections needed for rural or extended residential areas. Today, each household receives its electrical power via powerlines, it is connected via the good old copper wire with the telephone network, increasingly also via coaxial cable with the TV distribution network, and in future perhaps even via an optical fibre as "fibre to the home". Indeed, there are more than one seperate connections existing or possible, even today. But one should at this point not speculate on how the residential sector would be supplied with all kinds of services - it will come anyway after this problem has been resolved for the business centres and this is why the further debate should concentrate on these parts of the networks.

Mr. Kolbe raised the issue of different tariffication for business centres and remote and rural areas. While in Germany, and many other countries, the principle of uniform tariffication within a given territory (distance-dependent but independent from the location, whether business centre or rural area) is applied, one might imagine that - because of competition - tariffs between business centres would be lower than between remote areas although the distance would be the same in both cases. If one does not like this idea (possibly politicians), the question will have to be answered whether or not the taxpayer would have to subsidize the higher tariffs, because it is not at all obvious that the telecommunication company has to provide such subsidies.

In response to the question ***Mr. Brown*** confirmed that, in the United States, rural service is indeed more expensive than in big cities which is reflected by corresponding tariffs, including a "social" tariff ("life line") for low-income customers.

While admitting that he had not a perfect answer to the question of uniform tarifficiation, ***Mr. Roulet*** agreed that the trend will go towards cost-based tariffs, but that there will be "high cost" and "low cost" components for the same distance, and how this will be brought together will require a political decision for the EU or for the states themselves. For example, one might consider a certain tariff for basic services and, if agreed, thereafter comes the question on how this should be achieved, by the telecommunication company or by the taxpayer? He believes that such questions will come up in 1998 and it seems that politicians and customer organizations have not yet fully understood what will be the resulting consequences, a matter that would merit more reflection.

Mr. Rosenberg said he does not believe in subsidies as they may be cancelled and one faces afterwards the same situation as before. He does not want the fixed "state tariffs" but rather prefers "prices", like any enterprise does. But there are still technical hurdles to overcome; the technology of existing exchanges (pulse metering of calls) does not yet permit the desired price flexibility, e.g. to charge the call set-up (because that is expensive) and the holding of the connection separately. With modern digital exchanges which are now increasingly coming into service, much more price flexbility would be possible and, hence, customer friendliness, but it will still take time before it can be offered.

Mr. Tarjanne agreed on the current trend twoards cost-based tariffs but warned that there might be a change in the future. Who would, for instance, take care of the costs for connecting each school, each library, each hospital, etc. to the upcoming information highways now under discussion in the United States and elsewhere? He very much doubted that such costs would ever be covered by just letting the free market forces do its work, and there might be interesting developments affecting the tariff structures.

Mr. Chamoux quoted the example of the Swiss railway network in which the big cities are interconnected by the Federal Railway, while many small (even private) railways are providing service in the rural and mountain regions of Switzerland. He evoked that, in analogy, this Swiss model may possibly also be applied to telecommunication networks, i.e. to divest local networks to smaller private operators which would operate with different price structures than the wider and bigger global players whose costs might be higher because of more overheads.

In his reply, ***Mr. Ricke*** admitted that the local networks which are the most investment intensive part of the total network, generally do not cover their costs. Rather than handing them over to other operators he intended to turn them as much as possible into cost-covering parts. One should not forget that the local business is a high-volume market, representing a big share of the total turnover; for that reason, he would be thinking it over three times before handing over this business to someone else even if, for the time being, it is not profitable.

Summing up the discussion on uniform tariffication within a territory, ***Mr. Witte*** underlined that application of this principle does not mean it would automatically result in lower prices for each customer in general because the specific costs for additional service features would have to be taken into account. However, as evidence shows, in countries in which this principle has not been applied, prices are lower for customers who are using the basic telephone service without any additional service features.

Mr. Lorenz posed two questions to the panel: First, whether in view of the foreseeable developments, the telecommunication market - one of the few growing markets at all - will finally develop into a branch with meager profits and, second, whether the shareholders of Unisource will have in about 10 years united themselves into one single telecommunication company.

In his reply to the first question, ***Mr. Ricke*** stressed that the telecommunication companies will make every effort to stem the trend towards declining profits. But even if they cut costs and do their job well, it would also depend on the manufacturing industry and their efforts to offer competitive prices for their products. Much will also depend on the regulatory situation, what the regulations will look like under which the carriers will have to operate - and this is still an open question for the time being.

Referring to the second question, ***Mr. Rosenberg*** expressed his firm believe that the three telecom companies which are the shareholders of Unisource will still exist in ten years, although some of their present activities might well be outsourced to Unisource if it deems reasonable.

Mr. Krieger wanted to know from ***Mr. Brown*** whether the often-heard argument that deregulation leads to lower telecommunication coverage ("tele-density") had been confirmed the by the experience made in the United States. In a second question, addressed to ***Mr. Ricke*** and ***Mr. Roulet,*** he asked why there is opposition against participation of private capital and private companies? Why are there worries whether such companies (like Compagnie Général des Eaux) either make profits or loose money in telecommunications? Why worry whether there is only one network or whether there are five different networks if private capital is willing to finance them?

Mr. Brown confirmed that, in fact, the tele-density increased after deregulation; today, most homes have two or three telephone lines. Because of this high saturation in the United States, alternatives and lower prices are being offered by competition for the services in added value. Historically, in USA existed subsidized rates for farmers, for remote areas which had to be compensated by business, and metropolitan service. But now, the national policy has changed in restructuring prices and increased availability of services.

In his lively reply, ***Mr. Roulet*** stressed that he himself as Chief of France Telecom, is advocating further evolution of the present statute, including enforcement of capital influx. But in France, and also in Germany, decisions are to be taken by the Parliament. France Telecom is absolutely not against competition - this will definitely come sooner or later and cannot be avoided. Compagnie Général des Eaux is a private monopoly and without discussing what serves the general interest better, a private or a public monopoly, all what France Telecom requests are equal conditions for the competitors as ***Mr. Rosenberg*** already had emphasized. The second condition requested is a clearly forecasted timetable and a foreseeable context. If these two prerequisites would be fulfilled, he would be personally satisfied; with this basis one could develop suitable structures, synergies, market potentials, and so on - it would be a real challenge ahead.

Upon a short query of ***Mr. von Pattay*** concerning separated tariffication of the call-set-up and the holding of a connection, ***Mr. Rosenberg*** indicated that this might be possible around 1997.

Mr. Hultzsch recalled ***Mr. Witte's*** first question at the opening of the panel discussion, searching for reasons for "going global". He felt that the role of the customer (although briefly mentioned by ***Mr. Vuccins***) had been somewhat left out in the discussion. The customer will be a global information customer and the telecommunication companies will have to consider this when developing their services. This will provide broad competition and will help the telecommunication companies to stimulate and unfold their further growth.

In his closing remarks, ***Mr. Witte*** expressed his thanks to all speakers, Session Chairmen, and to the audience. While certainly not all aspects of the future telecommunication landscape could be covered, this first day of the Congress provided a wealth of information but has also identified many open questions which are waiting for an answer in view of the further evolution of the global telecommunication market place.

The Global Telecommunications Economy - Barriers to Entry

David M. Leive

I am going to talk to you today about barriers to entry in the global telecommunications economy, all sorts of barriers. But first, I need to sketch out what I mean by the "global telecommunications economy."

The concept as I see it has five overlapping components- services and equipment, players, market structures and telecom regulatory regimes. Let's briefly take up services and equipment first.

1 Components of the Global Telecom Economy - I

1.1 Services and Equipment

At the outset, I want to make several general points, to give some suggestion of the scale and magnitude involved.

Telecommunications appears to be growing at a much faster rate than other sectors of the economy. For example, international telephone traffic increased at a rate of 17% during the last decade.

Moreover, the telecommunications industry is very strong: more than 47 service providers and 15 equipment manufacturers have revenues greater that US$1 billion. They have an average level of pre-tax profitability of about 14%, which is much higher in some regions. Overall, there were world-wide generated revenues of US$1.8 thousand billion, and operating profits of US $200 billion in 1992 (12% of revenues).

Now to some specifics.

Services comprise 78% of the global telecommunications market of $535 billion, with equipment comprising the remaining 22%. And divided by region, the market for services is concentrated on the Americas and Europe, with a combined share of 75%, with Asia comprising most of the remaining 25%. See Figure 1.

Revenues from telephone services for nine selected countries are shown in Figure 2, along with their average annual growth rates. Growth rates for most western countries, with the exception of Spain, has been quite low. Comparing growth rate and relative share, for example: the very high growth rate for China, with its minuscule relative share.

Another way of looking at the situation is to examine main line density and main line augmentation. Figure 3 shows this data for 20 countries. There are roughly four categories. There are large, developed, but still growing markets -- like the

U.S., Germany and Japan. Some countries, such as Sweden and Switzerland, are already saturated, with high density and little augmentation. Others, like Brazil, Egypt and the Ukraine, have very low density and relatively low augmentation. Finally, countries like China and Russia have equally low densities but higher augmentation.

Figure 4 examines the same data, but in terms of revenues per main line. Note that in terms of revenue per main line, Mexico and Brazil may be more attractive than China, even though the augmentation in absolute terms may be much higher in China.

Why is this so? Several reasons suggested themselves as to the profitability of telecom operations in developing countries, although they may not explain the difference between Brazil and China. Usually, telecom services are provided under monopoly arrangements with little tariff control. Then, too, Government owned monopolies pay little or no taxes. Lastly, a high percentage of revenue comes from accounting rate payments from developed countries.

The growth in network digitization and specialized services and equipment has also been remarkable. Examples are the intelligent network markets, growth in cellular mobile telephone subscribers, public radio paging subscribers,and public radio paging subscribers. Figures 5 - 9 shows the data in terms of growth rates and absolute numbers.

Now let me turn from services to equipment.

Figure 10 shows the 1991 world market for telecommunications equipment, totalling about 82 billion U.S. dollars. Switching equipment constitutes about one third of the total, while, interestingly enough, satellites and cables comprise a relatively small percentage.

Let's take a closer look at the market for various types of terminals. Figure 11 shows what we all intuitively know: the tremendous rate of growth in the last decade, and the remarkable growth in the use of fax machines.

The world market for telephone terminals is shown in Figure 12, for fax machines (revenues) (Figure 13), for public switching equipment (Figure 14), and for private switching equipment (Figure 15).

The market for transmission equipment by type is shown in Figure 16. First, note the remarkable growth rate. Then, note the growth of the market for mobile communications equipment, largely at the expense of the metallic cables market, which experienced a steep decline. Figure 17 shows the market for mobile system equipment, with extremely strong growth rates in Europe.

1.2 Developing Countries

Let's spend a few moments looking at the particular problems of developing countries.

Many developing countries are characterized by large, unmet demands; and large segments of the population are without any services.

The marked global imbalance is shown in Figure 18. Developing countries (low income countries) accounted for 59% of global population but only 4% of telephone mainlines. The World Bank has defined low income countries with GDP per capita below US$600 per year; most are in Africa and Asia. At the same time, network growth in these countries is about 11% per year - about double the global average.

Telecommunications services in these low income countries are highly profitable, for reasons described above. Nevertheless, low income countries accounted for 17% of global demand but only 5% of global investment, as shown in Figure 19.

2 Components of the Global Economy - II

Let me now turn briefly to the three other components of the global telecommunications economy - the players, the market structures and the regulatory regimes.

2.1 The Players

The point to make is how radically the players have changed in the last decade or two. Fifteen or twenty years ago, the only players were the government-owned PTT's, or their functional equivalents, such as AT&T. Now, the great majority of these PTT's are either fully or partially privatized, or "corporatized", that is, organized into relatively autonomous state owned enterprises or entities. The data for the top 42 public telecommunications operators is shown in Figure 20. The breakdown of these 42 on extent of privatization is as follows:

- fully privatized: 17
- partially private: 7
- state owned enterprise: 13
- government department: 5

(Note that of the 17 fully private PTO's, 11 are from the U.S.)

Moreover, and again in contrast to the earlier situation, these privatized PTO's are not the only players in the game. Because of the new technologies, the pent-up demand, and the opening of formerly closed market structures, the new players include cable operators, cellular operators, VSAT systems operators, other types of wireless operators, variously funded by investment banks, commercial banks, equipment manufacturers and other entrepreneurs.

As remarkable as the multiplicity of new players is the way in which they are combining and alliancing with each other, as described by many of the other speakers at this conference. Among the key characteristics of these new players are their global or regional reach, their offering services that increasingly blur traditional service distinctions, and their practical need to cooperate with each other.

2.2 Market Structures

At one end of the spectrum (I hesitate to use the word "extreme"), are what I would call the "open" models, such as the U.S., with widespread competition, and no government ownership of facilities. At the other end of the spectrum are countries,, admittedly fewer and fewer, with a single monopoly PTT, little or no competition, and borders closed to external service provision and investment.

In between are the vast majority of countries, characterized by:

- privatized or corporatized PTTs
- varying amounts of competition,
  - in customer premises equipment
  - in cellular systems
  - in enhanced services, such as private line networks
  - in international voice service

Figure 21 provides some statistics for these liberalization trends. These statistics reveal, again, what we all know: that competition is greatest in the provision of telephone sets, that somewhat less than half of the 105 responding countries had some form of cellular competition (or at least the legal possibility of competition), but that only 10% of the responding countries permitted any form of international voice competition.

2.3 Regulatory Regimes

The final component of the global telecommunications economy, the telecom regulatory regimes, is also characterized by widespread diversity and an increasing, though not sufficient, rate of change.

Let me make some brief comments on three aspects of these regimes: context, regulatory functions, and organizational forms.

As to context, a country's telecom regulatory regime cannot be viewed in isolation. It needs to be evaluated in the context of the particular country's political system and institutions, as well as its political, economic and social objectives. It also needs to be viewed in the context of the structure of the existing telecommunications industry and the anticipated directions of change.

As to function, irrespective of its level of economic development and market structure, every country carries out some regulatory functions, even though they may not be characterized as such and government officials may vigorously deny that they are "regulating" anything. To make matters more complicated, these functions may be scattered amongst different ministries. The essential regulatory functions relate to licensing, interconnection, setting tariffs and prices, technical standards, type approval of customer premises equipment, and frequency spectrum matters.

As to organizational form, here too a range of different types exists. At one end, and perhaps here I will use the term "extreme", is the U.S. model, consisting of an independent regulatory agency with extremely detailed rules and procedures, and with complex interrelationships with other departments of the U.S. Government, with Congress and with state regulatory agencies. At the other end, and this is typical of many of the least developing countries, all regulatory and operational functions may be performed within a single ministry, without any separation of regulatory and operational functions, and with no clear rules or procedures.

In between these two extremes are the more typical cases, where the regulatory functions may be carried out by an identified entity or office within the Communications Ministry, with varying degrees of autonomy, resources and expertise.

This five part classification of the global telecommunication economy is, of course, not as neat and compartmentalized as my description may have suggested.

In reality, of course, the components overlap, are consistently affecting each other, and pull in different directions.

3 Barriers to Entry

With this background now out of the way, I would like to turn to identifying the barriers to entry and then examining what can be done to remove or reduce them. But first, some general comments, including some mention of several of the fallacies that seem to persist in this field.

It is perfectly clear to everyone, I believe, that technology, market forces and the new global players are overwhelming the existing governmental institutions for regulating telecommunications matters, both on the domestic and international levels. These regulatory regimes still reflect the simpler world of government monopoly providers, and monopolies do not need much regulation.

This brings me to the first of three fallacies, namely, that deregulation is equivalent to no regulation, or at least to much less regulation. It may superficially appear paradoxical, but it is nevertheless true, that deregulation often requires more rules and regulations, at least initially, to lay the groundwork for effective competition.

Let me give one example. A key ingredient to promoting competition is through an open entry policy. But to be effective, open entry requires interconnection to the public switched network, and the rules concerning such interconnection - the detailed terms and conditions and charges and technical requirements that will make it happen or not happen - can only be established by regulation.

The second fallacy is that privatization, by itself, is a panacea that somehow will assure competition. This is not so. Privatization assuredly will fill the governments' coffers, but unless it is simultaneously accompanied by provisions for new entrants, interconnection standards, and a revised regulatory regime with adequate authority over the privatized entity, a country may end up with perhaps the worst of both worlds: a private monopoly, no effective competition, and a limited regulatory ability to correct the situation.

The third fallacy is that the dividing line, for telecom regulatory purposes, is between the developed and developing countries. Many "developing" countries possess telecom regulatory arrangements and skills superior to those of some developed countries. Rather, for regulatory purposes, the dividing line does not lie between developed and developing countries - they form a continuum with many common problems - but between the twenty or so least developed countries possessing extremely limited resources and little or no formal legal or regulatory telecom structures, and all the others.

3.1 Barriers to Entry - A Catalog

I ought to start by describing what I mean by the term "barriers'. Rather arbitrarily, I have divided them into what I call "hard" barriers and "soft" barriers. "Hard" barriers comprise formal and explicit laws and regulations. "Soft" barriers relate to more intangible, but no less real, impediments, such as a lack of rules, inadequate personnel, lack of funds, obsolete practices. In fact, experience has shown that it may prove to be more difficult to change the "soft" barriers than the "hard" ones.

3.1.1 "HARD" Barriers

- No competition is permitted. This is perhaps the most obvious case. In a monopoly situation, neither domestic or foreign entrants can gain access to the country, so that there are total barriers to entry in terms of the provision of either facilities (equipment) or services, or both. The monopoly PTT is the only entity legally entitled to furnish equipment -- whether customer possesses equipment or transmission facilities, and is the only entity legally entitled to provide any communications service. While these barriers are coming down, the speed with which they are being dismantled varies greatly from country to country.
- Ownership restrictions. Even where a foreign carrier or other service provider is permitted to operator in the country, there are often restrictions relating to ownership, which limit the ability to joint venture with local firms. An example is the U.S. 20% limit of foreign (in revealing U.S. terms, "alien") ownership of US carriers.
- Trade restrictions. The barriers here relate to national efforts to protect domestic industries or domestic operators. This is a particularly sensitive issue in developing countries where the local industry may not be able to effectively compete in an open international procurement of telecommunications equipment. The recently concluded GATT accords should help in the removal of these types of barriers.

3.1.2 "SOFT" Barriers

These barriers are more intangible but nonetheless real. They comprise outdated telecommunications regulatory regimes on both the national and international level, even if such regimes do not contain any of the explicit "hard" barriers that I have just described.

Why should such regimes constitute barriers to entry?

These regimes are often characterized by an absence of substantive laws and regulations and an absence of procedures. A potential investor or service provider, whether foreign or domestic, may look in vain for some authoritative provisions concerning allowable rates of return, interconnection requirements, and the like. Often, there are no statutes, or the only applicable law may be, for example, a 1925 Decree that has never been translated, and whose existence is known to few people. Similarly, potential investors may look in vain for identifiable procedures and designated agencies and officials for particular matters.

It is for these reasons that the World Bank in a recent report set a top priority on regulatory reform - so that investors at least would know what were the rules of the game and how they were applied.

These "soft" barriers are encountered on the international level as well. For example, the rules and recommendations of the International Telecommunication Union regarding such matters as provision of service, access to facilities and services, and allocation of scarce national resources such as orbital locations and frequency spectrum, may be limited or unclear.

A couple of examples will illustrate my point.

National telecom regulators are only just beginning to grapple with the problems of how, if at all, to regulate the global or regional services provided by a global or regional venture or consortium. Does each country in which the service is provided impose its own patchwork of requirements, or should there be a harmonizing of licensing requirements? If so, by what process should this harmonization be done? Alternatively, should one consider a scheme, along the EC model, for mutual recognition of licenses issued by specified other countries?

This type of issue arises especially in connection with the new services that will be provided by low earth orbiting satellite systems, where services are provided on a global basis by a foreign entity, with the means of local control or regulation not clear and perhaps not feasible.

3.2 What to do About These Barriers?

How remove or reduce these carriers? How should open entry be promoted? How particularly should the soft barriers be handled?

Let me first address what role governments should play in ownership of telecommunications facilities, and then what role governments should play as regulators.

There need to be a continuing shift from government ownership and management of telecommunications facilities and services, so that its role is limited to policy and regulation. The increasing pace of privatization reflects this shift.

Governments also need to create conditions for the PTT to operate as a business, even where it has not been privatized. This means "corporatizing" the PTT, that is, separating it out from the Ministry so that it stands alone as a relatively autonomous entity, less subject to financial control, governmental personnel rules, and the normal political pressures that beset any Ministry.

What role should the government play as a Regulator?

The underlying fact is that, like any bureaucracy anywhere, regulators are adjusting slowly to the dramatic changes now taking place.

In trying to address this question, let's talk about two aspects: first, function, and, second, what I would term "style" or philosophy.

As to function, whatever a governments' style or philosophy, there are some irreducible regulatory functions, concerning the six matters I listed earlier: licensing, interconnection, tariffing, frequency spectrum matters, type approval of customer premises equipment, and standards.

As to style or philosophy, the issues can be framed by the following questions: How active should the regulator be? To what extent and in what areas should it defer to the marketplace? Should it take a pro-active role and create the right climate, or should it go further and try to "pick winners"? Or, on the other hand, should it simply "get out of the way? If so, in what areas?

In considering these questions, let me make a couple of preliminary points.

First, the regulator cannot be taken in isolation: it is necessary to take into account the role of the government as a whole, with its various ministries and agencies, and place the telecom regulator within that context.

Second, the term "regulator" seems clear enough, but actually embraces a wide range of different types of institutions, ranging from an FCC, to an OFTEL, to an autonomous unit within the Communications Ministry, to a single overworked official in a Ministry dealing with all regulatory and operational functions.

Lastly, the role of the regulator cannot be discussed in the abstract. Not only does it depend on the factors I just mentioned, but it depends as well on the particular country, its stage of development, and its own peculiar circumstances.

Having gotten these qualifications out of the way, let me hazard some specific suggestions.

First, whatever the philosophy, there needs to be clarity and transparency in the telecom regulatory regimes. Clarity, so that everyone, and particularly outside investors and operators, know what the rules of the game are. Transparency, so that everyone knows not only what the rules are, but how the game is played.

Second, regulators should facilitate competition and promote open entry, and encourage innovation in technology and services. Fine, but how?

Generally, by creating room for market forces to work, for example by reducing or eliminating licensing requirements for new entrants. Put another way, regulation should be "minimal" -- what is minimal cannot be decided in the abstract but the point is that there should be no automatic presumption in favor of regulation.

Perhaps it is easier to describe this in terms of what regulators should not do.

They should not guarantee success to any entrant or licensee - that is the function of the marketplace.

They should not normally seek to defend existing interests, such as privatized PTTs.

They need to be careful in a rapidly changing environment not to act prematurely: a careful balance needs to be struck: the regulatory entity needs to anticipate developments and not just deal reactively, but at the same time the transitory nature of regulation needs to be recognized: regulation seemingly current can become obsolete overnight.

And lastly, they ought to try and act in a technology-neutral fashion, and not try to pick a "winning" technology. This is a difficult enough task in any event, and is best left to the marketplace.

The global marketplace means that regulatory entities as a practical matter will possess less autonomy in deciding on some issues that need to be addressed - they will increasingly have to take into account issues on the regulatory agendas of other countries. This in turn suggests some efforts at harmonization, reciprocal licensing, and more organized efforts to exchange information among regulatory agencies.

Let me conclude this very general survey by returning to and reiterating my main theme -- that the barriers to entry in the global telecommunications economy are varied and pervasive, that it will take time to dismantle or change them, particularly those more amorphous barriers based on custom and attitude, but that this task needs to be done, and sooner rather than later, if the benefits of technology and market forces are to be widely available.

The Global Telecommunications market

Share of the telecommunications market, by sector and by region, 1992

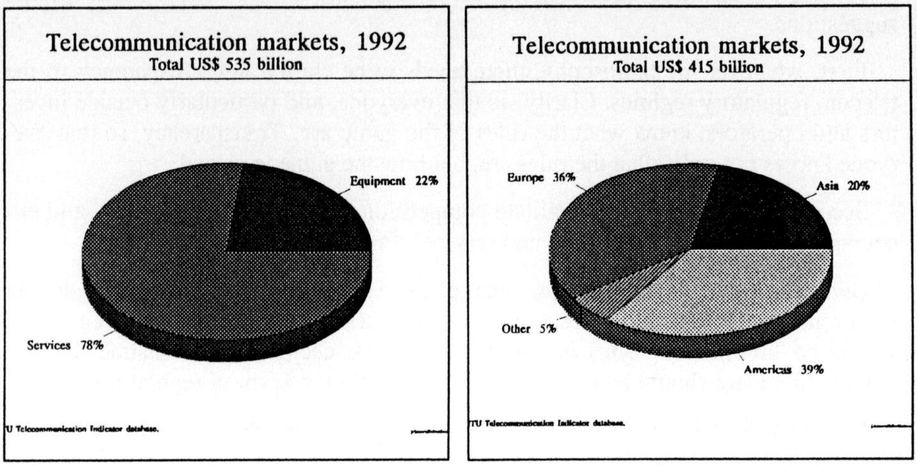

Figure 1

Revenues from telephone service

	1987	1988	1989	1990	1991	%	relative share 1991
Canada	11,50	12,00	12,00	12,30	12,40	1,90%	4,40%
China	1,10	1,30	1,60	1,80	2,00	16,12%	0,70%
France	15,90	14,00	14,40	15,80	16,60	1,08%	5,90%
Germany	23,40	23,30	23,50	23,60	23,90	0,53%	8,50%
Italy	14,30	15,00	15,70	15,90	16,70	3,95%	5,90%
Japan	37,40	39,00	38,90	39,20	39,40	1,31%	13,90%
Spain	6,40	6,90	7,40	8,20	8,90	8,59%	3,10%
UK	15,90	15,90	16,10	16,50	17,00	1,69%	6,00%
USA	120,20	117,10	111,80	107,50	104,80	-3,37%	37,10%
Others	30,90	32,50	36,40	56,20	40,30	6,87%	14,50%
Total	277,00	277,00	277,80	297,00	282,00	0,45%	100,00%

billions of 1991 fixed value dollars
%: average annual growth rate

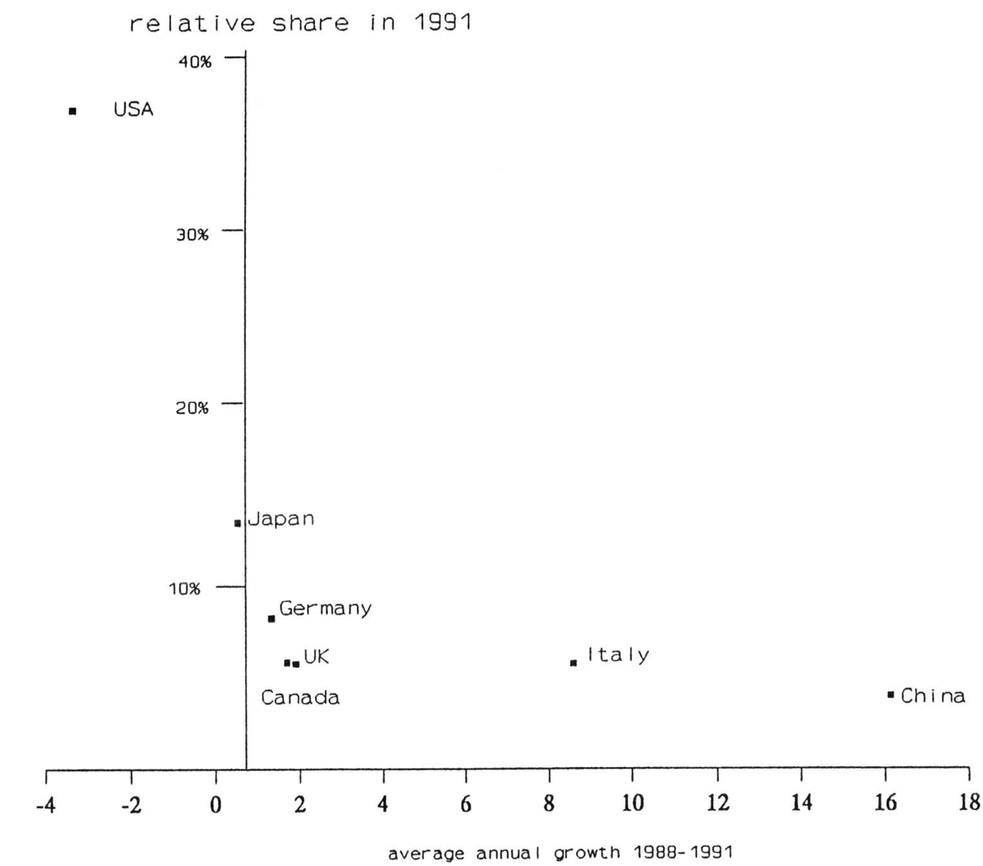

Figure 2

Main Lines: Density and Augmentation

Figure 3

	Augmentation	Density
USA	3321	55,27
Japan	1730	46,06
Germany	1672	41,81
China(PR)	1600	0,72
Russia	1393	13,75
Korea(Rep.)	1318	34,45
France	960	50,15
Italy	720	39,93
Brazil	666	6,57
Spain	661	33,99
Mexico	951	7,06
Canada	519	58,52
Ukraine	466	12,98
Poland	272	9,29
UK	270	45,74
Greece	241	40,80
Egypt	237	3,65
Switzerland	137	59,48
Sweden	99	69,00

Augmentation: Main lines added in 1991, in thousands
Density: Main lines per 100 inhabitants, 1.1.1992

Source: SIEMENS

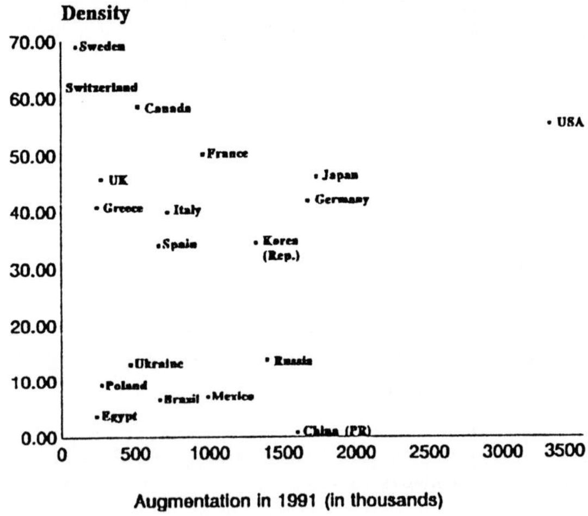

There are roughly four groups of countries

- The large, developed, but still growing markets like the USA, Japan or Germany (upper right). - In Germany, the unification resulted in a reduced overall density (about minus 10%) and an extraordinary augmentation in main lines.

- The smaller markets in industrialized countries, e.g., Switzerland, Canada or Sweden with high density rates and comparatively low augmentation (upper left).

- A third group consists of countries with low density rates and rather low augmentation like Poland, the Ukraine, or Brazil.

- Finally, countries like China or Russia show a considerable augmentation in main lines. Considering the revenues per main line, however, a market like Brazil looks far more interesting than e.g., China.

Main Lines: Density and Revenues

	1986	1987	1988	1989	1990	1991	%	Density
Brazil	376	422	425	550	562	631	10,91%	6,57
Canada	894	854	862	819	819	807	-2,03%	58,52
China	128	139	155	166	169	173	6,21%	0.72
France	664	641	542	536	561	573	-2,90%	50,15
Germany	840	851	820	800	768	752	-2,19%	41,81
Hungary	455	481	427	429	353	308	-7,51%	10.69
Italy	757	747	747	738	725	732	-0,67%	39,93
Japan	759	759	763	731	715	696	-1,72%	46.06
Mexico	480	518	541	611	599	576	3,71%	7,06
Spain	592	629	633	624	653	622	0,99%	33,99
UK	710	727	696	672	660	657	-1,54%	45,74
USA	1100	1021	958	888	821	774	-6,79%	55,27

Revenues per main line in 1991 fixed value dollars
%: average annual growth rate
Density: Main lines per 100 inhabitants, 1.1.1992

Sources: OMSYC, SIEMENS

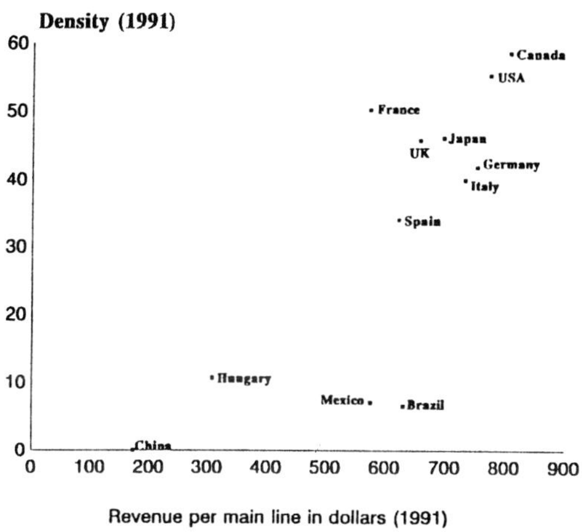

Figure 4

Network Digitisation of Subscriber and Trunk Lines in OECD Countries 1991/92

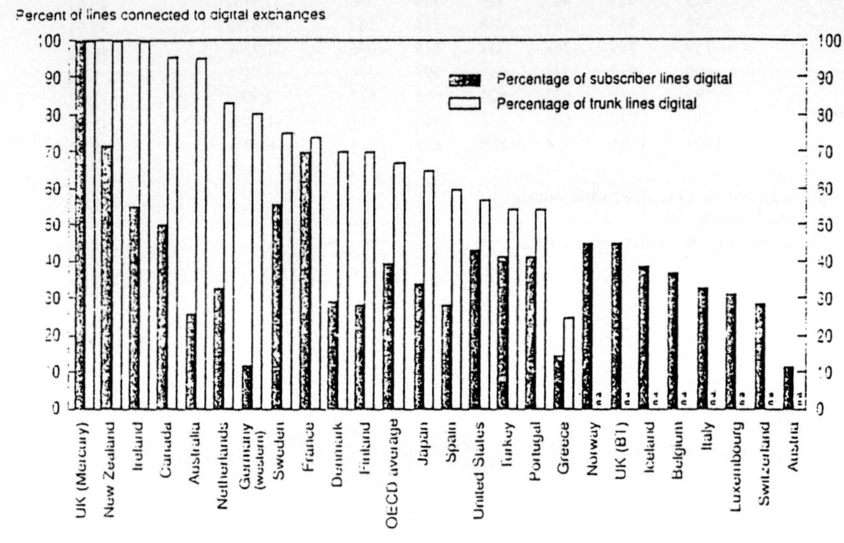

Source: OECD

Figure 5

Intelligent Network Markets

	Equipment	Services
France	543	1303
Germany	186	1215
UK	480	1404
USA	10095	20414

Source: Communications Week, 7 February 1994

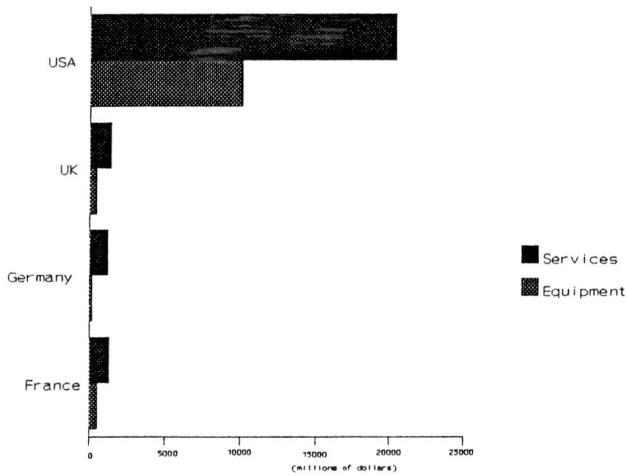

Presently the German market for intelligent networking is comparatively small. According to London consultancy OVUM Ltd., Germany's spending on IN equipment will outstrip the UK's in 1997, however, with a projected amount of $675 million. Germany is considered to become the biggest IN market in Europe.

Figure 6

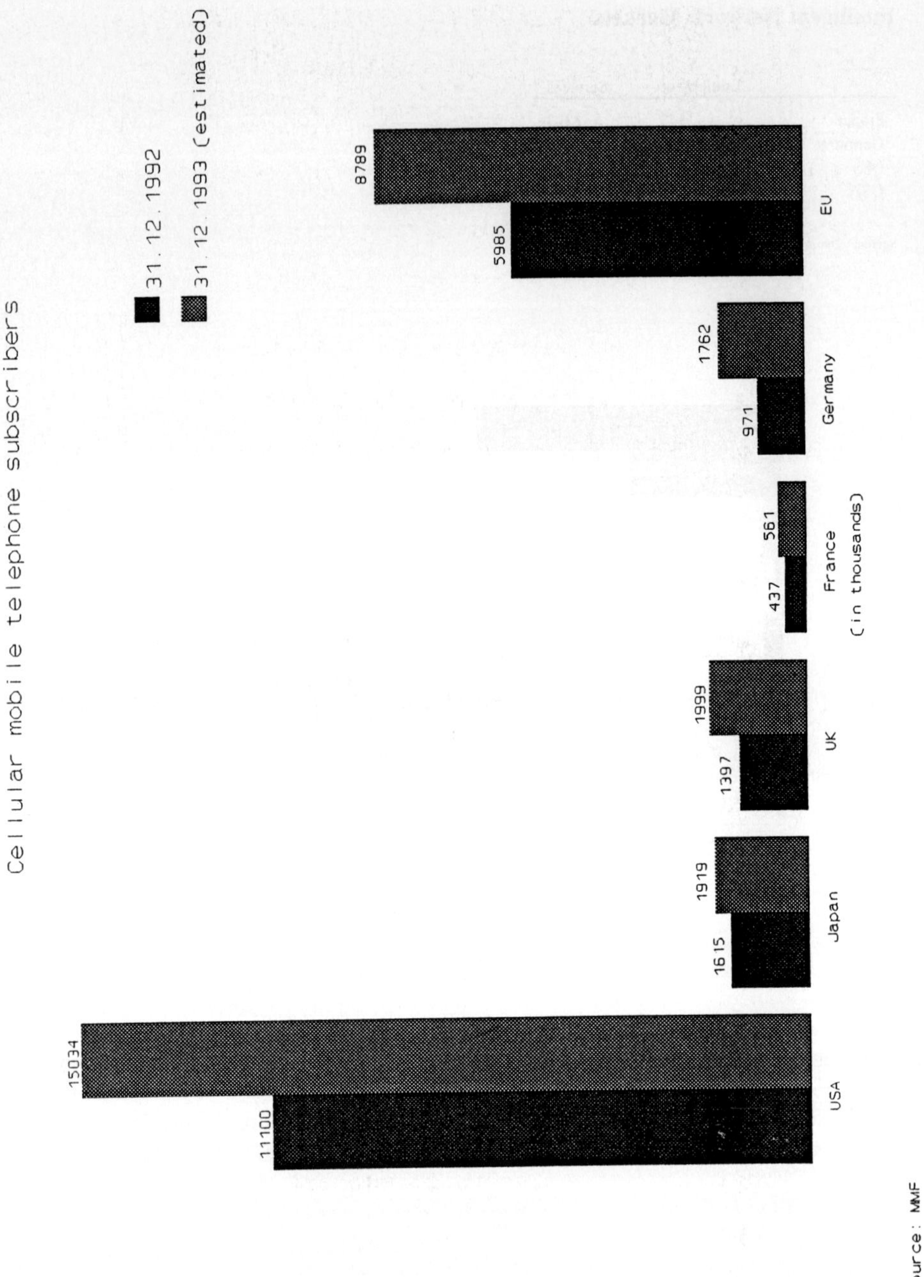
Cellular mobile telephone subscribers
31.12.1992
31.12.1993 (estimated)
11100
15034
USA
1615
1919
Japan
1397
1999
UK
437
561
France
(in thousands)
971
1762
Germany
5985
8789
EU
Source: MMF

Figure 7

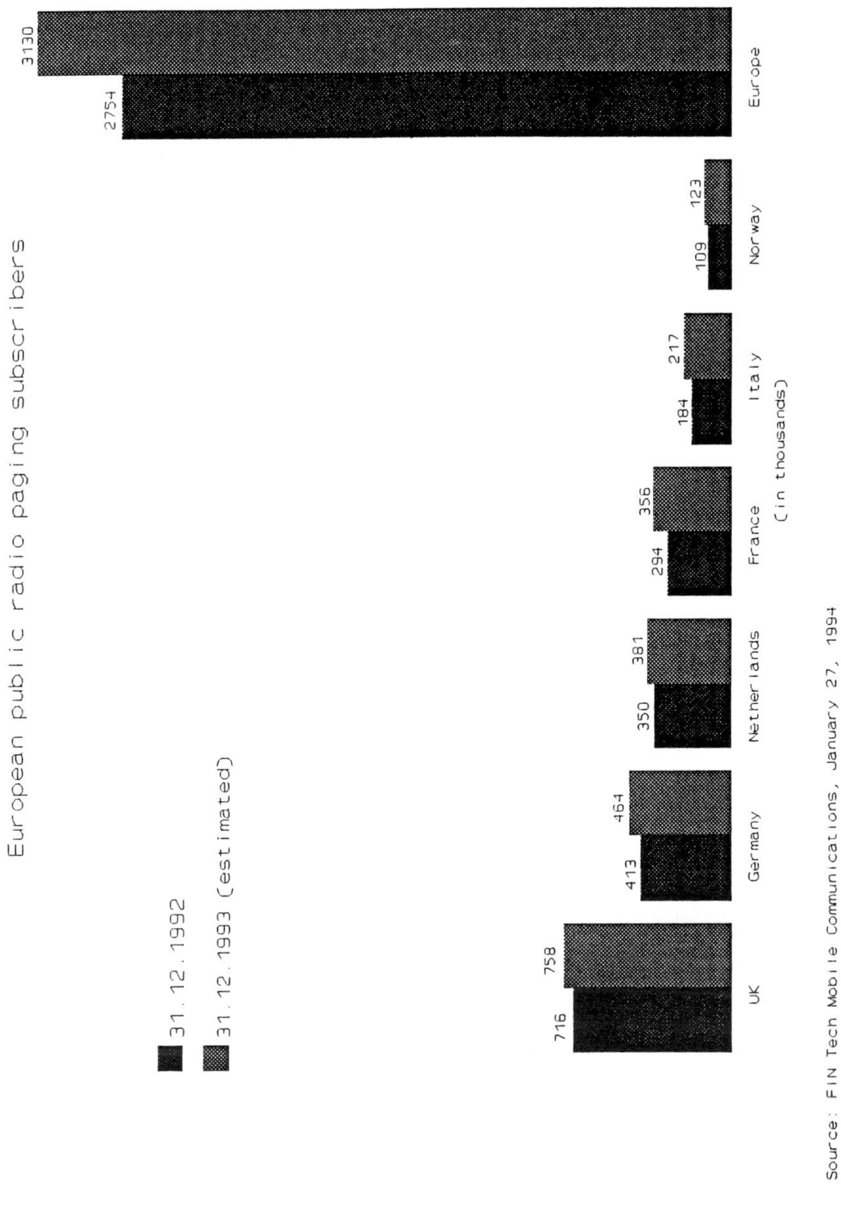
European public radio paging subscribers
31.12.1992
31.12.1993 (estimated)
716
758
UK
413
464
Germany
350
381
Netherlands
294
356
France
184
217
Italy
109
123
Norway
2754
3130
Europe
(in thousands)
Source: FIN Tech Mobile Communications, January 27, 1994

Figure 8

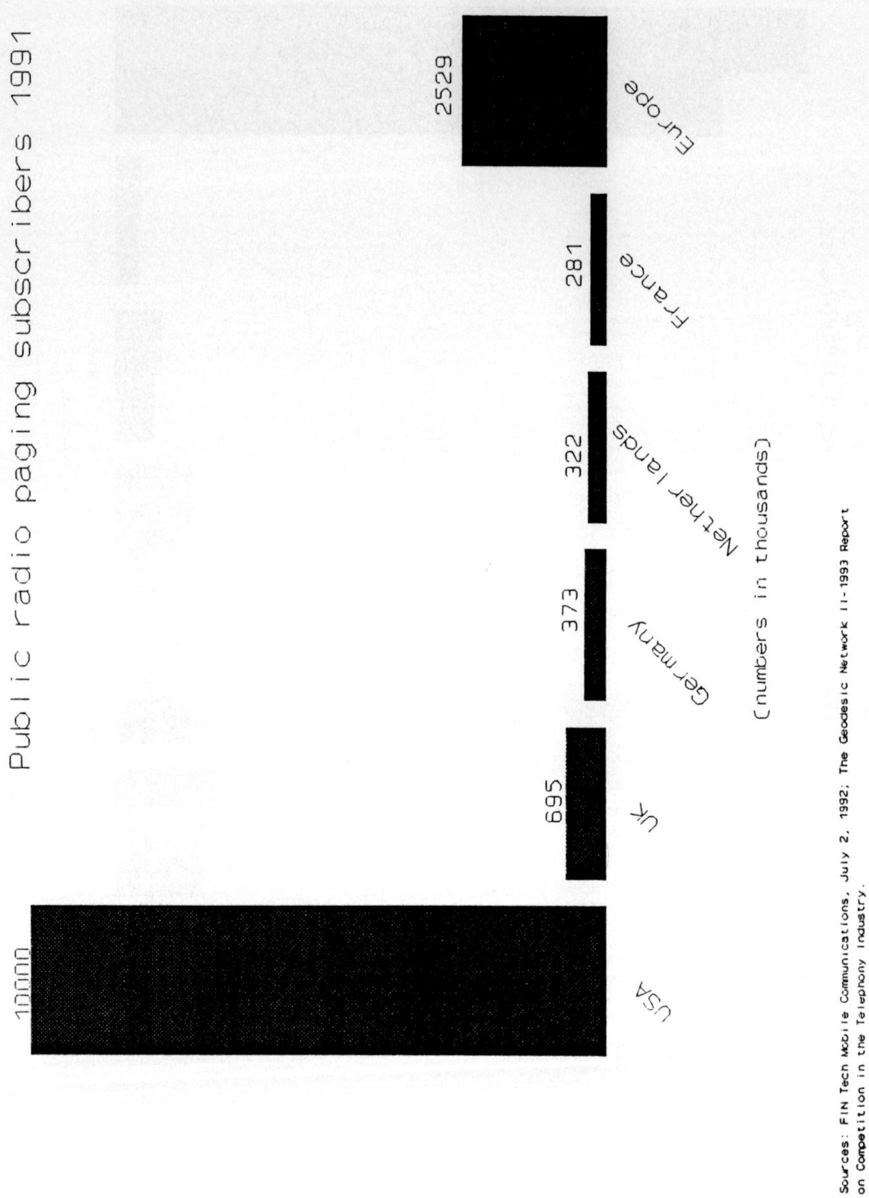
Public radio paging subscribers 1991
10000
USA
695
UK
373
Germany
322
Netherlands
281
France
2529
Europe
(numbers in thousands)
Sources: FIN Tech Mobile Communications, July 2, 1992; The Geodesic Network II-1993 Report on Competition in the Telephony Industry.

Figure 9

1991 World Market of Telecommunication Equipment

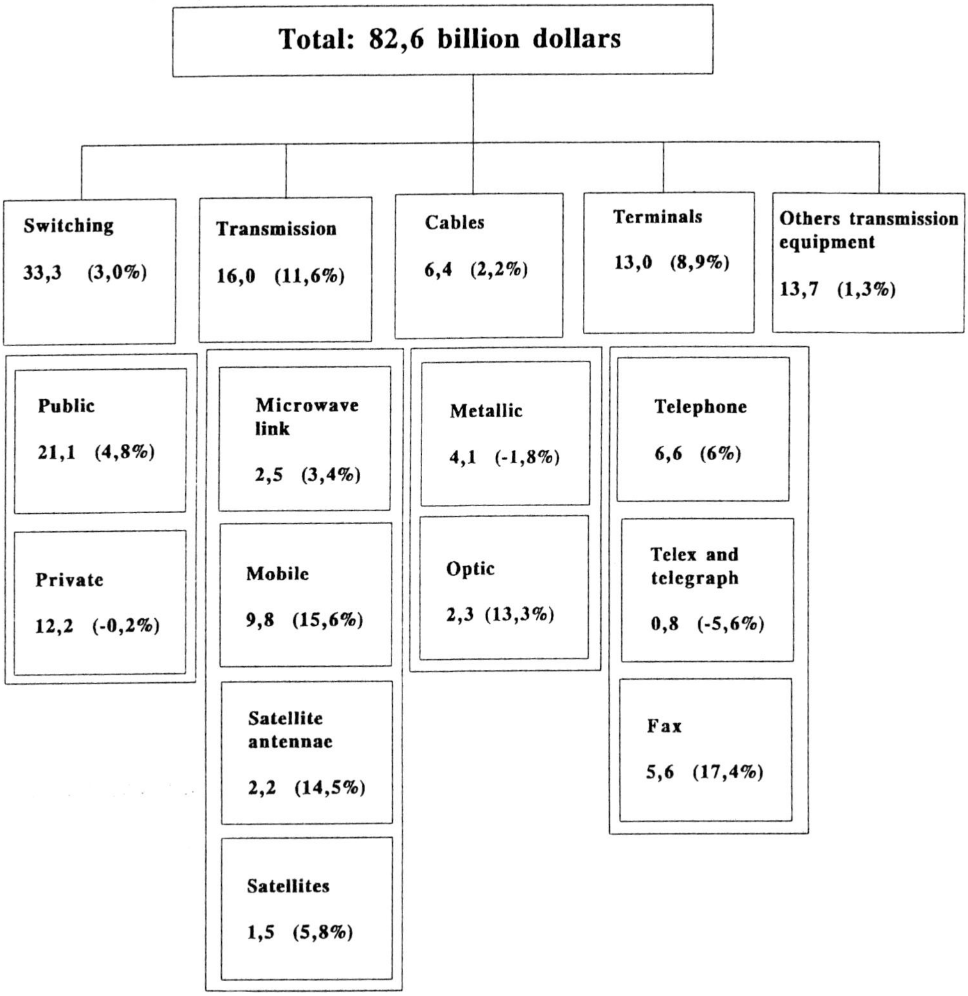

Figure 10

World market for terminals by type

	1986	1991	%
Telephone terminals	4,99	6,67	6,00%
Telex terminals	1,03	0,77	-5,60%
Fax machines	2,49	5,57	17,40%
Total	8,51	13,01	9,00%

billions of 1991 fixed value dollars
%: average annual growth rate

Source: OMSYC

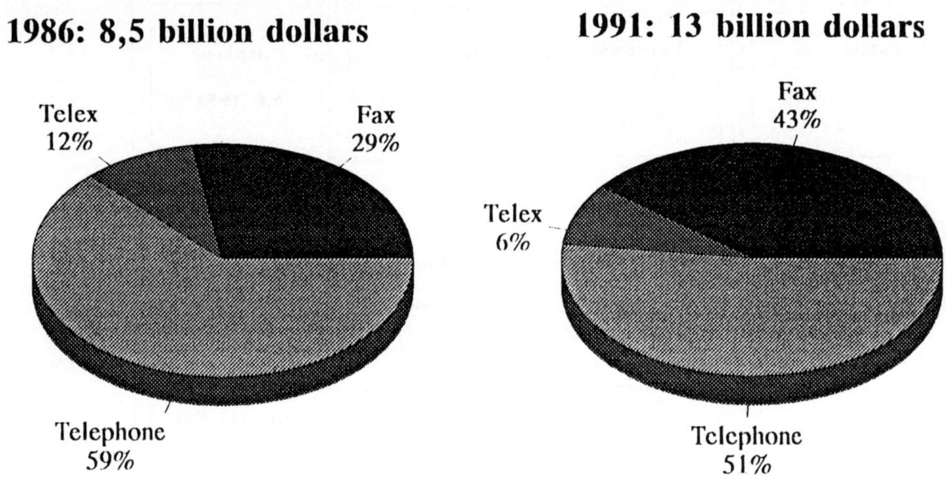

Figure 11

World market for telephone terminals

	1987	1988	1989	1990	1991	%	relative share 1991
Canada	0,18	0,18	0,18	0,19	0,19	1,36%	2,80%
France	0,34	0,39	0,39	0,39	0,40	4,15%	6,00%
Germany	0,61	0,66	0,70	0,71	0,71	3,87%	10,70%
Italy	0,19	0,20	0,21	0,22	0,24	6,01%	3,60%
Japan	0,73	0,98	1,14	1,18	1,20	13,23%	18,00%
Spain	0,18	0,19	0,19	0,18	0,18	0,00%	2,70%
UK	0,33	0,31	0,30	0,29	0,29	-3,18%	4,30%
USA	1,73	1,75	1,76	1,73	1,73	0,00%	25,90%
Others	1,35	1,51	1,60	1,66	1,73	6,40%	25,90%
Total	5,64	6,17	6,47	6,55	6,67	4,28%	100%

billions of 1991 fixed value dollars
%: average annual growth rate

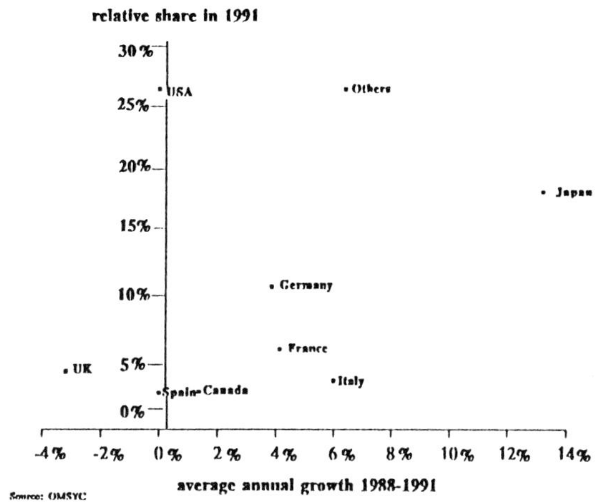

The telephone terminals market comprises subscriber telepone sets (including wireless telephones) and answering machines. --The trends observed are influenced by strong price effects.

Figure 12

World market for fax machines (revenues)

	1988	1989	1990	1991	%	relative share 1991
Canada	126	136	147	157	5,65%	2,80%
France	189	244	266	300	12,24%	5,40%
Germany	248	341	362	384	15,52%	6,90%
Italy	177	223	257	294	13,53%	5,30%
Japan	975	925	926	916	-1,55%	16,50%
Spain	91	137	144	163	15,69%	2,90%
UK	331	335	348	370	2,82%	6,70%
USA	1242	1345	1331	1347	2,05%	24,20%
Others	1020	1269	1440	1636	12,54%	29,40%
Total	4399	4955	5221	5567	6,06%	100%

millions of 1991 fixed value dollars
%: average annual growth rate

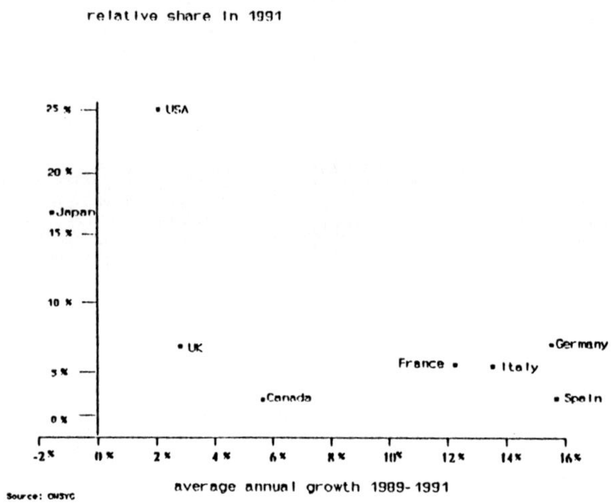

During the last years the world market for fax machines experienced a considerable growth. Comparing the growth in numbers of installed fax machines with the growth in revenue generated, the sharp price effect becomes obvious.

Figure 13

World market for public switching equipment

	1987	1988	1989	1990	1991	%	relative share 1991
Canada	0,65	0,67	0,67	0,69	0,70	1,87%	3,30%
France	0,96	0,87	0,77	0,74	0,73	-6,62%	3,50%
Germany	1,09	0,95	0,98	0,97	0,97	-2,87%	4,60%
Italy	0,86	0,95	1,03	1,06	1,08	5,86%	5,10%
Japan	1,91	2,33	2,78	2,75	2,68	8,84%	12,70%
Spain	0,36	0,65	0,83	0,85	0,93	26,78%	4,40%
UK	1,48	1,63	1,57	1,49	1,45	-0,51%	6,80%
USA	6,79	6,77	6,60	6,40	6,36	-1,62%	30,00%
Others	3,78	5,02	5,66	5,96	6,27	13,49%	29,60%
Total	17,88	19,84	20,89	20,91	21,17	4,31%	100,00%

billions of 1991 fixed value dollars
%: average annual growth rate

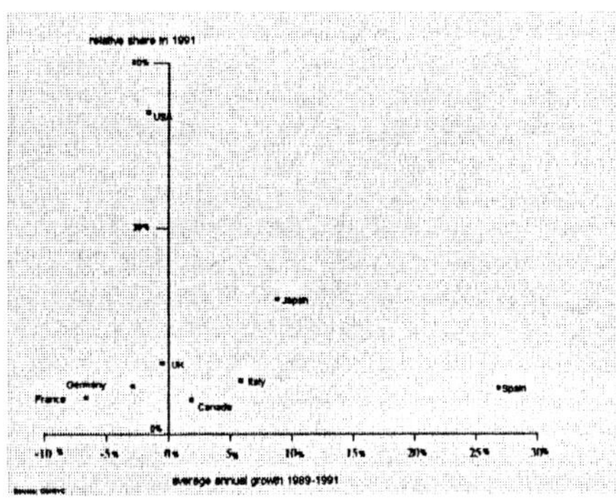

In 1991, public switching accounted for 63,5% of the world market for switching equipment compared to 36,5% for private switching. – In spite of a slow negative growth, the USA remain the largest single market. Spain has experienced strong growth during the last years, its relative market share surpassing already France.

Figure 14

World market for private switching

	1987	1988	1989	1990	1991	%	relative share 1991
Canada	0,38	0,37	0,37	0,38	0,38	0,00%	3,10%
France	0,43	0,46	0,44	0,45	0,47	2,25%	3,90%
Germany	0,98	1,07	1,00	0,98	0,97	-0,26%	7,90%
Italy	0,35	0,32	0,31	0,33	0,34	-0,72%	2,80%
Japan	1,30	1,37	1,34	1,39	1,42	2,23%	11,70%
Spain	0,17	0,24	0,26	0,26	0,28	13,29%	2,30%
UK	0,83	0,87	0,86	0,83	0,82	-0,30%	6,80%
USA	6,31	5,99	5,32	5,06	4,92	-6,03%	40,30%
Others	2,04	2,39	2,47	2,53	2,59	6,15%	21,30%
Total	12,79	13,08	12,37	12,21	12,19	-1,19%	100%

billion of 1991 fixed value dollars
%: average annual growth rate

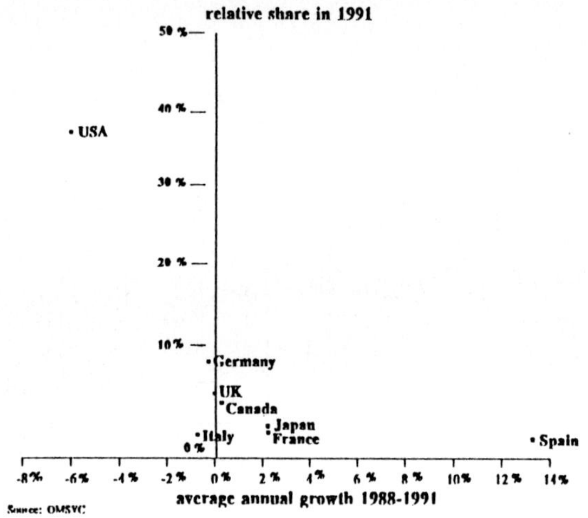

The market for private switching equipment includes PABX and intercommunication systems. The revenue generated there decline somewhat during the last years (at an average of -1,19%). As this is a very competitive market, price effects certainly are an important reason.

Figure 15

World market for transmission equipment by type

	1986	1991
Optic fibres	1,24	2,30
Metallic cables	4,56	4,10
Microwave links	2,13	2,52
Satellites	1,14	1,52
Antennae	1,12	2,20
Communication with mobile stations	4,75	9,83
Total	14,94	22,47

(billions of 1991 fixed value dollars)

1986: 14,97 billion dollars **1991: 22,47 billion dollars**

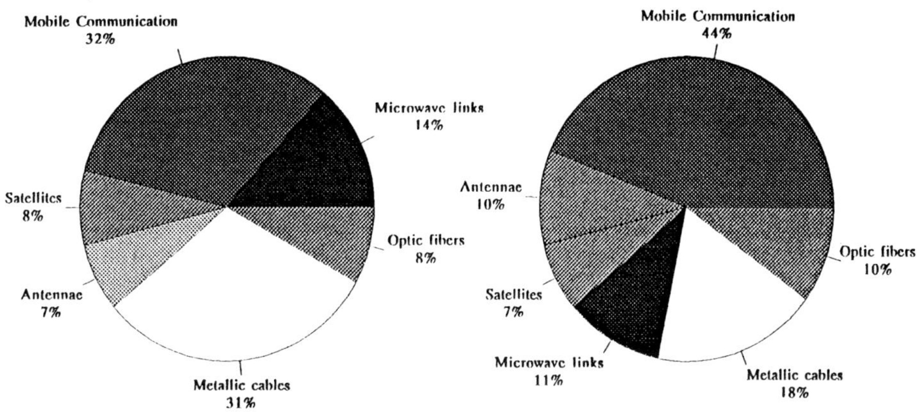

Figure 16

World market for communication equipment with mobile stations

	1987	1988	1989	1990	1991	%	relative share 1991
Canada	0,19	0,24	0,26	0,30	0,33	14,80%	3,40%
France	0,30	0,41	0,44	0,48	0,53	15,29%	5,40%
Germany	0,49	0,61	0,63	0,75	0,87	15,43%	8,80%
Italy	0,15	0,16	0,23	0,27	0,31	19,90%	3,10%
Japan	0,62	0,76	0,86	0,87	0,88	9,15%	9,00%
Spain	0,05	0,08	0,11	0,17	0,24	48,02%	2,50%
UK	0,63	0,88	0,89	0,88	0,89	9,02%	9,10%
USA	2,36	2,91	3,07	3.09	3,18	7,74%	32,40%
Others	1,44	1,87	2,16	2,36	2,58	15,69%	26,30%
Total	6,23	7,92	8,65	9,17	9,81	12,02%	100,00%

billions of 1991 fixed value dollars
%: average annual growth rate

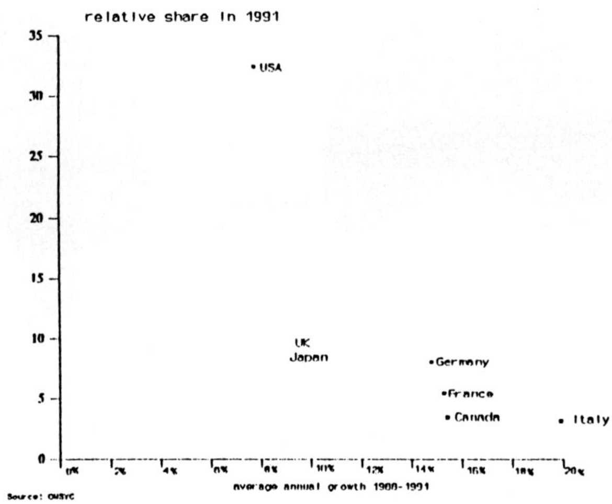

The market for equipment with mobile stations includes private systems and terminals as well as cellular telephone and paging eqiupment. The build-up of new networks has stimulated strong growth in Europe. The US market is still growing, slower though on a higher level.

Figure 17

In 1992, low income countries accounted for:

59% of global population ...

World population Total 5'463m

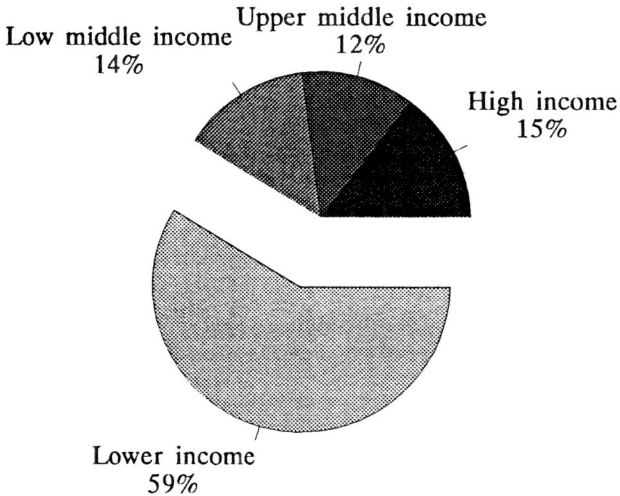

Telephone main lines Total 573m

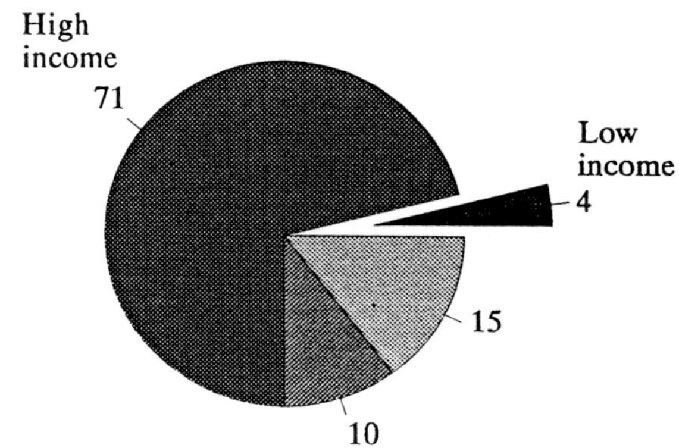

... but only 4% of telephone mainlines

SOURCE: ITU

Figure 18

In 1992, low income countries accounted for:

17% of global demand ...

Demand 1992, total 75m lines

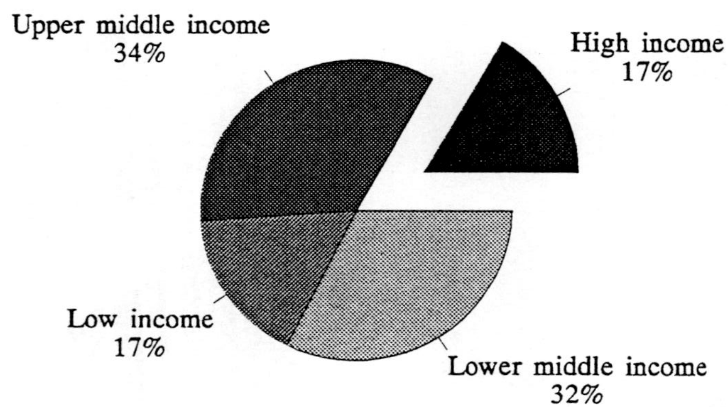

Investment 1992, total US$120bn

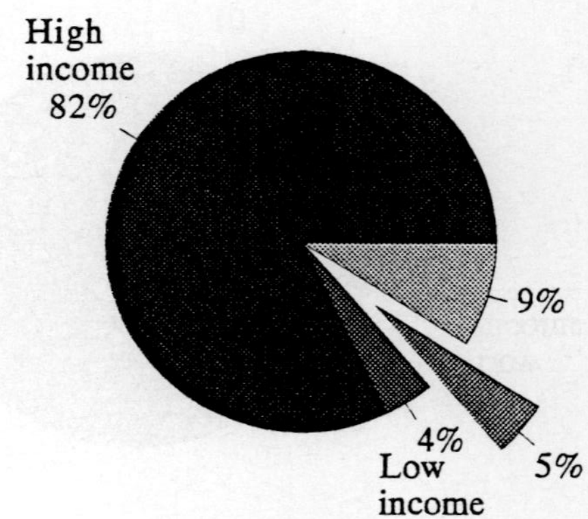

Note: "Demand" = new lines plus official waiting list

... but only 5% of glóbal investment

SOURCE: ITU

Figure 19

PUBLIC TELECOMMUNICATIONS OPERATORS WITH 1992 REVENUE IN EXCESS OF US$1.5 BILLION

Rank	Operator & Country		Revenue		Mainlines		Employees		Ownership
			Total (US$M)	Change, 1991-92 (%)	Total (M)	Change, 1991-92 (%)	Total (k)	Change, 1991-92 (%)	(Legal Status)
1	NTT	JPN	51,354	1.7%	57,300	2.7%	242.3	-6.0%	PP
2	AT&T	USA	39,580	2.0%	n.a.	n.a.	77.9	-0.5%	FP
3	DBP Telekom	DEU	34,550	14.3%	35,421	5.6%	231.0	0.9%	SOE
4	BT	GBR	23,379	-0.7%	26,084	1.9%	170.7	18.9%	FP
5	France Télécom	FRA	23,164	5.9%	30,100	3.4%	155.3	-0.5%	SOE
6	SIP	ITA	17,492	10.8%	23,709	2.8%	87.5	-0.1%	PP
7	BellSouth	USA	15,202	5.2%	18,677	3.4%	97.1	1.1%	FP
8	NYNEX	USA	13,155	0.6%	15,699	1.9%	70.9	-1.8%	FP
9	GTE	USA	12,644	0.6%	16,819	3.6%	81.1	-4.7%	FP
10	Bell Atlantic	USA	12,093	2.5%	18,181	2.4%	71.4	-7.2%	FP
11	Telefónica	ESP	11,279	14.5%	13,792	4.0%	74.4	-1.4%	PP
12	AMERITECH	USA	11,153	3.1%	17,001	2.5%	71.3	-3.6%	FP
13	MCI	USA	10,562	11.3%	n.a.	n.a.	31.0	11.2%	FP
14	US West	USA	10,281	-2.8%	13,345	3.2%	63.7	-3.2%	FP
15	SW Bell	USA	10,015	7.3%	12,803	3.3%	59.5	-2.8%	FP
16	PacTel	USA	9,935	0.4%	14,551	2.0%	61.3	-1.4%	FP
17	Sprint	USA	9,230	5.1%	n.a.	n.a.	43.4	0.5%	FP
18	Telstra	AUS	8,992	28.3%	8,257	2.6%	70.8	-12.7%	SOE
19	Bell Canada	CAN	6,505	1.7%	9,229	2.3%	52.9	-3.2%	FP
20	TELMEX	MEX	6,175	15.8%	6,546	12.1%	48.9	-1.1%	PP
21	PTT Telecom	CHE	6,012	14.3%	4,185	2.6%	19.5	0.8%	SOE
22	KT	KOR	5,985	13.4%	15,593	7.0%	59.1	2.1%	SOE
23	PTT Telecom	NLD	5,964	8.0%	7,395	3.1%	34.9	1.4%	SOE
24	TELEBRAS	BRA	5,900	25.7%	9,900	8.1%	89.6	0.6%	SOE
25	Telia AB	SWE	5,363	0.5%	5,919	0.1%	34.1	-12.7%	SOE
26	China P&T	CHN	4,570	42.9%	11,469	35.7%	559.1	5.2%	GD
27	Taiwan DGT	TWN	3,954	15.0%	7,137	4.1%	36.8	0.3%	GD
28	Telkom	ZAF	3,727	--	3,524	--	62.0	-8.4%	SOE
29	BELGACOM	BEL	3,203	6.9%	4,264	4.1%	26.8	-3.1%	SOE
30	PTV	AUT	3,083	5.5%	3,466	3.7%	18.3	-0.1%	GD
31	HongKong Telecom	HKG	2,795	17.8%	2,820	6.7%	15.9	2.8%	FP
32	TeleDenmark	DNK	2,582	1.9%	3,003	1.8%	17.7	-1.3%	PP
33	TUR-PTT	TUR	2,542	86.4%	9,472	16.2%	88.1	-2.9%	GD
34	Tele	NOR	2,440	6.2%	2,268	3.2%	14.8	-2.4%	SOE
35	Mercury (1)	GBR	2,117	31.0%	143	--	--	--	FP
36	KDD	JPN	1,927	-0.2%	n.a.	n.a.	5.9	-2.0%	FP
37	MOPTT	SAU	1,833	4.0%	1,568	7.0%	20.6	1.0%	GD
38	BEZEQ	ISR	1,703	6.0%	1,845	8.3%	10.0	2.4%	SOE
39	SGP Telecom	SGP	1,694	20.5%	1,101	6.2%	10.9	10.9%	PP
40	B.C. Tel	CAN	1,685	5.2%	2,080	4.6%	14.5	-3.3%	FP
41	TASA	ARG	1,598	43.7%	2,008	12.7%	19.3	6.3%	PP
42	OTE	GRC	1,584	23.1%	4,497	7.3%	26.7	-3.2%	SOE
	Top 42		**409,001**	**82.5%**	**441,242**	**5.3%**	**3,047**	**-4.9%**	

Note: 1. 1991 main line data; Shows major domestic units rather than ultimate owners to reduce double-counting.
Key: FP=Fully Private; PP=Partially Private; SOE=State Owned Enterprise; GD=Government Department; n.a.=not applicable.
Source: ITU Telecommunication Indicator Database.

Figure 20

LIBERALISATION TRENDS

Responses to the 1992/3 Liberalisation Questionnaire of the ITU Development Sector

	International Voice Competition			Cellular Competition			Telephone Set Competition		
REGION	Y	N	P	Y	N	P	Y	N	P
Africa	0	27	0	5	20	2	13	9	5
Americas	5*	20	0	15	9	1	21	3	1
Asia	2	16	2	7	12	1	13	5	2
Europe	1	22	2	10	10	5	23	2	0
Oceania	1	3	0	1	3	0	2	2	0
ex-USSR	0	3	1	2	2	0	2	1	1
Total	**9**	**91**	**5**	**40**	**56**	**9**	**74**	**22**	**9**

Key: Y = Yes, N = No Competition, P = Partial Competition.
Note: * The replies refer to the legal possibility of competition rather than the existing situation. Replies received from 105 countries.
Source: ITU Development Sector Liberalisation Questionnaire.

Figure 21

MCI: Creating One World Network

Eugene Eidenberg

Part I: Opening Remarks

A. Thank yous to Dr. Vitte and colleagues.

B. Theme set-up.

1. The international market is growing rapidly.

    a) Transborder volumes increasing at more than 20 percent annually; even faster in aggressively developing countries such as Mexico.

    b) Dramatically outpacing domestic growth in many countries.

2. Coupled with customer demand for greater choices in switched services, bandwidth on demand, emerging interactive multimedia services.

3. Burgeoning need for a network of networks capable of delivering on a global scale.

4. Key variables to developing such a network are:

    - Customers/Market.
    - Regulatory Environment.
    - Technology.

5. MCI — having been instrumental in developing the U.S. model for competitive networks — is now, with its alliances, a

major player with our partner BT in creating a world network of networks to supply the emerging global economy.

6 Today's Talk.

a) First, look at the U.S. model — although not perfect — to provide insights into how MCI, and all of us, will be affected by telecoms' key variables in creating one world network.

b) Second, look at NewCo, our joint venture company with BT, and why we believe this will be the global model for telecommunications one-stop shopping.

Part II: Customers — What do they want?

A. Proliferation of customers all over the world — Eastern Europe, Pacific Rim, developing Central and South American countries, in addition to already established and growing markets of North America, Western Europe and Japan. Increasingly, our customer markets are regional or global — not national.

B. All companies are behind the curve when it comes to customers. The customers are moving faster, finding new uses for, and demanding more from telecommunications than any of the companies can provide alone.

C. Customers want choice; choice gives them greater control over their businesses, their lives, etc.

D. U.S. model has taught us that customers prefer the ability to choose between a variety of players for:

1. Price.

2. Value.

3. Product feature/functionality.

4. Customer service.

5. Networks that can do what's important to them.

6. Billing.

E. Expanding into a global telecommunications environment customers will also need:

1. Consistent standards— regardless of location.

2. Competitive cost structures.

3. Consistent levels of service — redundancy — of switched multinational networks.

4. Strong presence in each country.

5. Solution flexibility.

6. Worldwide billing and local currency options.

Part III: Regulatory Environment

A. By these examples and our U.S. model we know that by opening up markets to competition everyone benefits.

1. Markets grow faster.

2. Customers have opportunity to get services they need.

3. Telecommunication carriers more profitable.

B. AT&T story: By losing market share, MCI's principal competitor grew more rapidly and without competition.

C. However, to truly achieve all benefits a fair and competitive climate must be welcomed on a global basis.

1. Telecom has risen to the trade agenda — services agenda in GATT.

2. Transborder network deployments regionally and globally will face a transition, overtime, away from accounting rates system to a model which looks like retail models — cost based, with the necessary relationship to rates.

Part IV: Technology

A. Vendor independence allows for "best of breed" products and network capabilities to be developed.

B. Technology alone won't be the competitive edge.

C. Technology exists today, but truly global network must meet needs:

1. Intelligent network platforms.

2. Seamless and borderless solutions.

3. Voice and data integration.

4. More locations.

5. Bandwidth increase.

6. Frame Relay.

7. LAN integration.

8. Videoconferencing.

Part V: NewCo — The MCI-BT Joint Venture

A. Partnering is smart strategy.

1. Stentor, Banacci (in markets rapidly approaching a liberalized, competitive environment).

2. Wraps-up North American continent with MCI intelligent network and shared strength of alliance partners.

B. NewCo answers the challenge of a global economy: Redefining the word "service" and offering world-class standards of performance.

1. Provides scale to support and service customers worldwide.

2. Provides broad range of uniform, "best of breed" market offerings.

3. Collaborates on development, deployment and operation of consistent, end-to-end, integrated managed platform.

4. Controls costs through managed public network with broad product appeal.

5. Focuses on speed to market and added value products.

Part VI: Conclusion

A. networkMCI — long-range strategic vision to be a global player and provider of services customers want. A string of initiatives to ensure we can deliver on the expectations of the global network. Creating a network of networks.

B. International Information Superhighways.

1. Evolving networkMCI.

2. Evolving VP Gore's national information infrastructure to international scale.

C. One world network.

1. Remain focused on the customer.

2. Expedite a regulatory environment that permits what customers need.

3. Continue to drive innovative technology.

4. We create a network for which time zones, languages and currencies are irrelevant, and provide service that seems local, even when it is being provided from the other side of the world.

5. Thank you.

Part VII: Question and Answer Session

Global Services and Partnership Strategies Goals

Ph... Walker

Global Services and Partnership Strategy Goals

21 April 1994

Sprint

A National Partnership Model

A commercial or equity relationship between Global Carrier and an incumbent or emerging PTO

- **Which provides PTO with:**
  - Family of seamless global voice and data services and resources
  - Right to use global brands (or co-brand)
  - Product platforms and business systems for use in-country and with global offerings
  - Global resources and partners to address customer needs
  - Additional traffic beyond bilateral streams
- **Which provides Global Carrier with:**
  - Compatible in-country networks and service offerings
  - In-country marketing, sales, and support organizations
  - Licensees for Global Carrier's intellectual property
  - Additional traffic beyond bilateral streams

Sprint

Telecom Trends

- Users, particularly multinationals, are demanding global services
- Technological developments are decreasing costs
- International tariffs are decreasing
- Privatization and liberalization are occuring worldwide
- Smaller carriers are emerging to compete with incumbent players
- Shifting from traditional bilateral relationships to global consortia
- Global Carriers and Consortia will target
  - Corporations
  - Consumers
  - National Carriers

Sprint

Customer Requirements

- Global connectivity
- Uniform, seamless, worldwide functionality and performance
- End-to-end provision of a full array of services and systems, guaranteed by one party (single supplier)
- Lower costs for all services
- Offer of packages of services tailored and flexible to meet customer applications
- Worldwide support services
- Billing and tracking systems which integrate network information on a regional and global basis
- Outsourcing to minimize customer management of non-core resources, to avoid non-strategic capital spending and simplify global network management

Sprint

Regulatory Environment

There are several worldwide telecommunications regulatory and policy trends, including;

- Privatization of established carriers in developed and developing countries;
- Liberalization of value-added services and telecom equipment;
- Liberalization of the use of leased lines, including interconnection to the PSTN
- Liberalization of the provision of switched services; and
- De-monopolization of telecoms infrastructure.

Sprint

New Industry Patterns

- New telecom technologies are creating alternative carriers and access networks. Such emerging carriers may need "global carriers" to deliver their global traffic.
- Traditional carriers now compete aggressively in international services by providing "carrier select" services
- New technologies are creating economic and political pressures for more open international arrangements.
- New "global" network services (Iridium) proposed by Motorola, will also lead to new "interconnection arrangements".

Sprint

International Market Sectors

- Switched voice services to other countries
- Transit services for carriers
- Transmission capacity for carriers

→ GLOBAL CARRIER

- Voice, video and data service
- Store & forward services
- Network Management
- Outsourcing

→ GLOBAL BUSINESS NETWORK

- Calling card products
- Country to country calling
- Platforms

→ GLOBAL BRAND

Sprint

International Carrier Services

Transmission Facilities

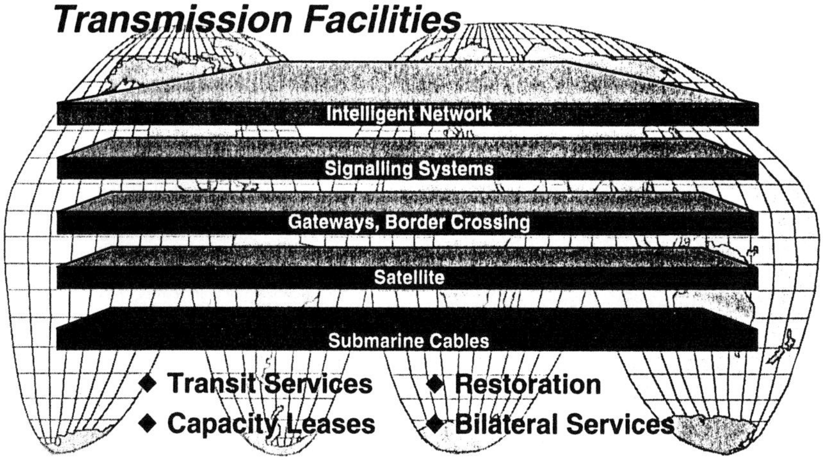

- Transit Services
- Capacity Leases
- Restoration
- Bilateral Services

"Carrier's Carrier"

Sprint

International Network Solutions

Global Network

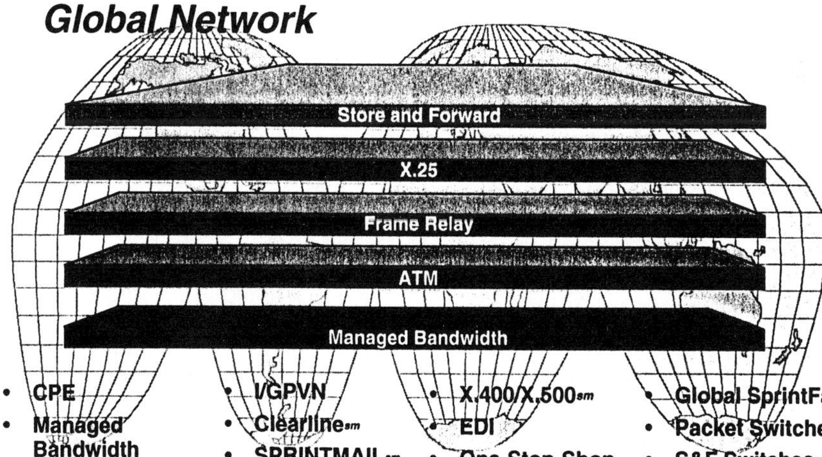

- CPE
- Managed Bandwidth
- TRANXACT

- I/GPVN
- Clearline[sm]
- SPRINTMAIL[sm]
- Customized Solutions

- X.400/X.500[sm]
- EDI
- One-Stop-Shop
- Hybrid Networks

- Global SprintFax
- Packet Switches
- S&F Switches

Sprint

International Consumer Market
Global Cashless Calling Services
Information Services
Calling Card Services
Multilingual Operator Services
IDDD
• International Toll-Free Services
• Visaphone
• The Most
• Global Calling
• Sprint Express
• FONCARD
Sprint

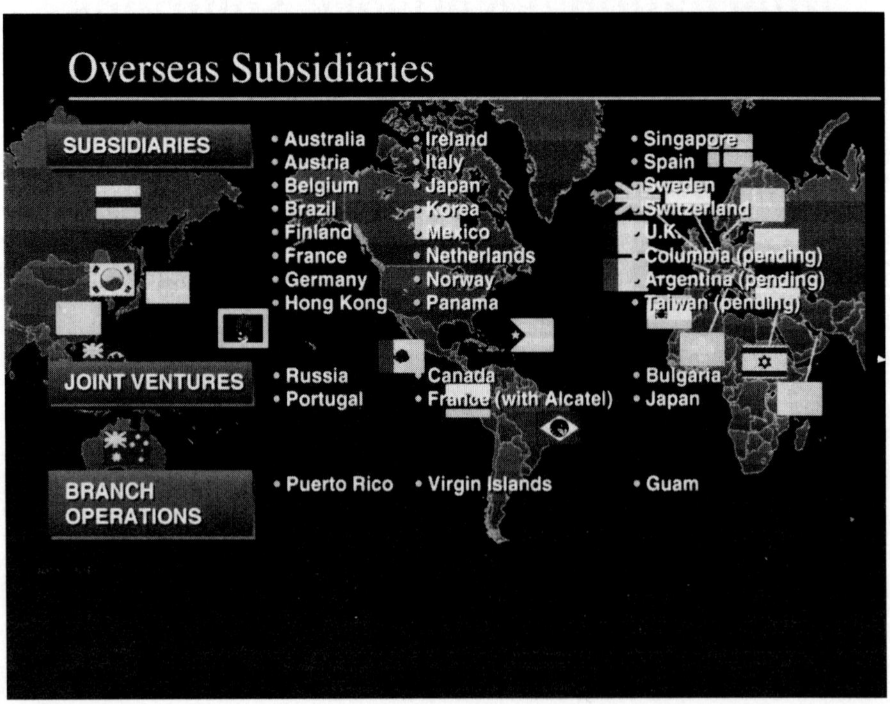
Overseas Subsidiaries
SUBSIDIARIES
• Australia
• Austria
• Belgium
• Brazil
• Finland
• France
• Germany
• Hong Kong
• Ireland
• Italy
• Japan
• Korea
• Mexico
• Netherlands
• Norway
• Panama
• Singapore
• Spain
• Sweden
• Switzerland
• U.K.
• Columbia (pending)
• Argentina (pending)
• Taiwan (pending)
JOINT VENTURES
• Russia
• Portugal
• Canada
• France (with Alcatel)
• Bulgaria
• Japan
BRANCH OPERATIONS
• Puerto Rico
• Virgin Islands
• Guam

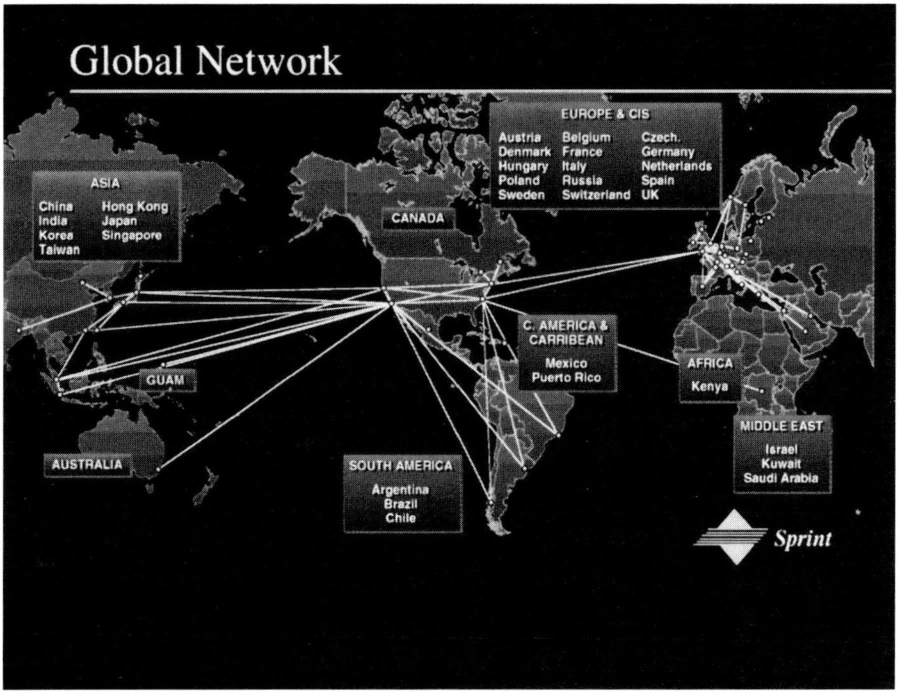
Global Network
EUROPE & CIS
Austria Belgium Czech.
Denmark France Germany
Hungary Italy Netherlands
Poland Russia Spain
Sweden Switzerland UK
ASIA
China Hong Kong
India Japan
Korea Singapore
Taiwan
CANADA
C. AMERICA & CARRIBEAN
Mexico
Puerto Rico
AFRICA
Kenya
GUAM
MIDDLE EAST
Israel
Kuwait
Saudi Arabia
AUSTRALIA
SOUTH AMERICA
Argentina
Brazil
Chile
Sprint

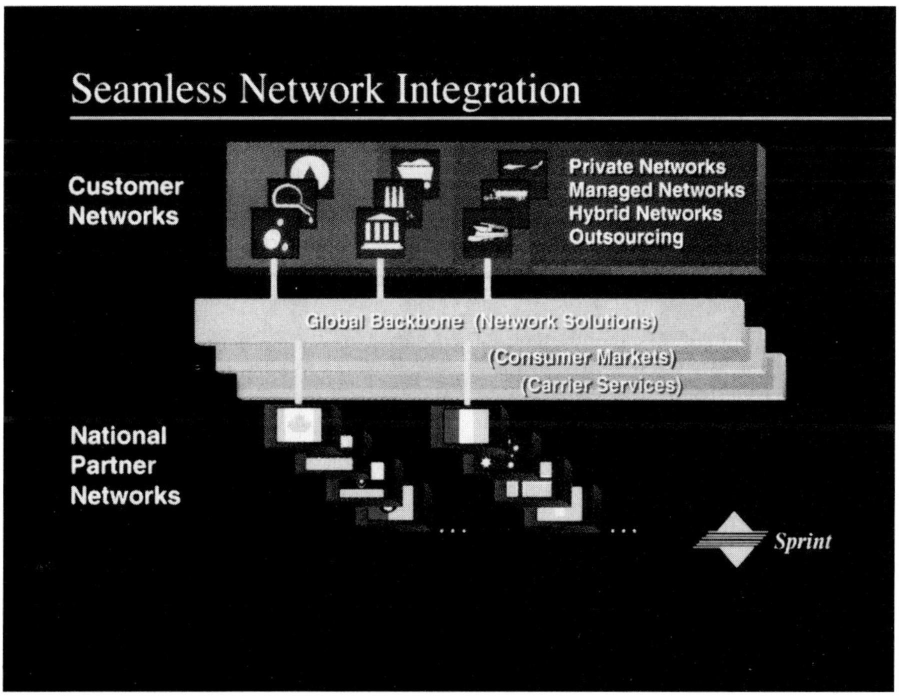
Seamless Network Integration
Customer Networks
Private Networks
Managed Networks
Hybrid Networks
Outsourcing
Global Backbone (Network Solutions)
(Consumer Markets)
(Carrier Services)
National Partner Networks
Sprint

A Global Partnership Model

- ◆ Collaboration of like-minded carriers
- ◆ True global orientation and commitment
- ◆ Accelerated marketing of global products
- ◆ Pooling of resources for investment
- ◆ Transport economies of scale
- ◆ Positioning for pan-European and other regional networks
- ◆ Sprint roles criteria for recognizing partners

Sprint

A National Partnership Model

Criteria for recognizing partners

◆ **Incumbent PTO candidate**
- Need for global services and competitive intellectual property
- Threatened in own market
- Likely to affiliate with global or regional coalition

◆ **Emerging carrier candidate**
- Capital and influence
- Concession and/or rights of way
- Customers and sales capabilities
- Existing network facilities
- Ambition to become a full service operator

Sprint

Bellsouth Strategies for Global Participation

Charles B. Coe

Good afternoon. On behalf of my company, I want to convey our thanks to Dr. Witte for giving BellSouth the opportunity to participate in this prestigious gathering.

BellSouth has been involved in the global telecommuni-cations market since 1987 and we have maintained a consistent commitment to increase our participation as the global market has expanded with opportunities to participate. In that time, we have been fortunate enough to become one of the industry's leaders in the provision of wireless communications and the development of alternative networks.

I would like to take a few minutes to establish just why the opportunity to participate in this congress is appreciated and why the topic "Global Players in Telecommunications" is so important to a company that is frequently and anachronistically, in my opinion, referred to as an RBOC - i.e., a Regional Bell Operating Company.

BellSouth is currently licensed to provide a wide variety of telecommunication services to customers in 14 countries on 5 continents. Outside the U.S., we have participated in the building and operation of businesses that offer mobile data, long distance, CATV, paging, business network and cellular services. In cellular, the businesses in which we have an interest serve over 800.000 customers overseas and encompass license areas of over 200 million POPS.

Cellular, in fact, is the most rapidly growing segment of our business overall. In the U.S., BellSouth's growth exceeded 45 percent in 1993 with our two millionth subscriber added last month. Outside the U.S., our cellular businesses are experiencing an even greater growth rate - 250 percent in 1993. Much of this phenomenal growth is attributable to normal startup growth curves. However, there are other factors which have contributed that I will touch upon later.

On a basis proportional to our equity ownership, BellSouth currently has more than 250,000 cellular subscribers outside of the U.S. Total revenues from our non-U.S. business, again proportional to our equity, will exceed U.S. $550 million this calendar year and will reach U.S. $1 billion by 1996.

So, while the financial scope of our non-U.S. business is still small relative to the U.S. business, it is by far BellSouth's largest and most important growth activity.

It would be understandable to conclude that BellSouth's business is still predominately confined to the U.S. domestic market, or even more narrowly, to the regional U.S. domestic market. However, I strongly suggest that such a conclusion is misleading with respect to both the present and the future of BellSouth.

On that note, I will now turn to the topic of my presentation, which is "BellSouth's Strategies for Global Participation". While I will talk about strategies that have worked and are workig for us, I don't expect them to be remarkably different from the strategies that have worked for other successful operators who have pursued border-crossing expansion of their business.

There is, however, one element of our strategy that I will describe more fully because I believe that it is very important to the industry and very much in the interest of all industry players, whether they be traditional monopolists in a given market, or whether they be new entrants to a given market. That element is to support and foster certain key principles of liberalization and competition.

Before explaining this, I want first to identify and discuss other important elements of BellSouth's strategy for global participation.

This industry has had to cope with the issues of partnerships and alliances more intensely, and in a briefer period of time, than any other industry with which I am familiar.

With few exceptions, I think that the industry players have done a good job at working with each other and with new industry entrants in a manner which I would describe as situational cooperation - that is, where politics, market conditions or the rules of competition dictate a partnership. Describing these successes as situational cooperation is not intended to be pejorative but, rather, is intended to point out that when cooperation is required, the successful global players have been able to do it and do it well.

At BellSouth, we would certainly like to think of ourselves in that category. We set out as a strategic principle that we will be a good and effective partner. In each of our partnerships, we try to achieve an alignment of interests and values, as well as an align-ment of strengths and competencies in a complementary manner.

What we offer, in addition to the more apparent attributes, is an absolute commitment to integrity as a partner. I believe it is important in a partnership to promise only what can be delivered, and to deliver all that is promised.

Another feature of our strategy initially is to have our partners, over time, think of us as being a little - well - predictable. Now, I do not mean predictable in every sense of that word. When it comes to performance, technology and vision, we intend to be growing and changing. But, at the same time, we try to be entirely transparent as to our interests and direction - we are, therefore, reliable to our partners, and, in that narrow sense, predictable.

Situational cooperation has brought the industry to a new level of maturity and perhaps, even, a new level of necessity. We, as well as many other players, are now examining our strategy for partnerships to determine if it is time for a strategy of strategic partnership - in other words, a partnership or partnerships that reach beyond the expediency of individual opportunities.

If the press has it right, then every company represented in this room is working on regional or global alliances. I suspect that for all of us, these partnership decisions are going to be enormously more difficult, and the alignment of interests and vision significantly more challenging, than the partnerships of situational cooperation.

Certainly the successes of the past give us some encouragement about the prospects for the future. Presumably, if one can be a partner in a discrete business, then those skills should be transferable to partnerships that are more broadly based.

However, the new skill set is probably incomplete. BellSouth's strategy for the future will be to emphasize and develop the skills necessary for the new era of partnership.

Our second critical strategy for global participation is what I would summarize as a strategy of focus. Of course, every company will claim that they are focused. But, the proof lies in the record. And, I submit that the record demonstrates that some companies have used their strategy for focus more effectively than others. For the seven years that BellSouth has been a global participant, our strategy for focus has included:

• A geographical focus - we have emphasized Western Europe, Latin America, Australia and selected countries of Asia for our development activities. At present, our non-U.S. investments - which amount to more than U.S. $ 1 billion - are almost equally divided among our Europe and Latin America and Asia-Pacific Regions. Increased participation in Europe holds the highest priority for BellSouth. We are in cellular partnerships in Denmark, Germany and France, in the mobile data partnerships in the U.K., France, Belgium and the Netherlands. Further, we expect to be active in every GSM and Mobile Data license announced in the EU for 1994. We have been bolstered by the increasing number and quality of the market openings in Europe, and our proven ability to contribute in harmony with European partners. In many respects, we see that both the market liberalization and BellSouth's participation are still in their early days in Europe.

• As another matter of focus, we have shown a preference for start-ups over acquisitions as the principle manner to grow our participation in overseas businesses.

• And, with more than 80 percent of our investment devoted to cellular, we have demonstrated a clear commitment to wireless technologies and services. Only recently have we begun to expand our activities into fixed network development.

For example, in Australia, our company has both a GSM license and a general carrier license. There we have built and are operating a fixed long distance and international network that includes direct connections to business customers. Also, in Chile, we are nearing completion of a fixed network to enter the long distance and international calling businesses as logical extension of our already successful cellular business.

I will certainly admit that maintaining this strategic focus has been difficult. There have been many opportunities outside the parameters we have set that have been particularly alluring. And, in retrospect, we might even wish that we had pursued some of them. One example of this is some of the Telco privatizations in Latin America, a region where we have solidly established ourselves. In the end, our decisions to pass on opportunities outside of our chosen focus were not based on opportunity attractiveness. These decisions were instead based on the strategic imperative not spread our resources and expertise so thinly that we jeopardized the quality of our performance.

No doubt, as we continue to grow in our global participation, our strategy of focus will be redefined, when appropriate, to encompass our increased capacities and interests.

Complementary to focus, it has been our strategy to emphasize BellSouth's core competencies. Among those I would list end-user marketing, especially mass marketing, customer service techniques, network design and network project management.

All of our successes have been achieved by applying these core competencies in concert with the core competencies of partners. To be a core competency, we believe that skills and performance have to attain the status of "global best practices". That accounts for my relatively narrow view of what our core competencies are. Clearly, we do a lot of other things but none better than those I have mentioned.

Further, I attribute a significant measure of the growth in our business, which I mentioned earlier, to the successful application of core competencies. Nowhere is that more evident than in Venezuela, where we have experienced profits and positive cash since the first year of operation. We currently have 130,000 customers in Venezuela, and these are customers who use their cellular phones three times as much as our U.S. customers!

The fourth strategy we have followed will initially sound more like an issue of style or execution than a strategy. However, from the perspective and context of BellSouth's development as a global participant, the early, explicit decision to follow a strategy of broadbased corporate commitment to build and sustain success, was one of the most important strategic decisions we have made.

While I suggested early that it is misleading to think of BellSouth today as a regional company, that undeniably was our focus in the beginning. In the early days, few of our managers even had passports... even fewer had any business experience outside of the U.S. And most of that experience was in the area of procurement, as we were both sought out and courted by prospective new suppliers following our split from A&T.

Therefore, we had little practical experience to draw upon in our move towards global participation. However, we were strongly motivated by both a vision that technology and global economics were going to dramatically change the old order of telecommunications, and by a keen desire to grow our business. This early vision has repeatedly been confirmed by industry developments with which we are all familiar. And, the desire to grow globally has been reinforced by the ever-increasing competitive assaults on our traditional markets.

Motivation is a good start, but we recognized, early, that success would be dependent upon a consistent and disciplined commitment to transform ourselves from a regional to a global player. I would like to cite some of the ways in which this corporate commitment has been demonstrated and continues to be central to our overall strategy.

First, there is commitment to investing for the long term. As previously noted, we have emphasized startups which, as you know in the network infrastructure business, are characterized by several years of losses and relatively long paybacks, even when highly successful.

To undertake this, particularly on an appreciable scale, is difficult for any earnings-oriented company, especially an American one whose quarterly results are the largest single factor in its share price. To endure dilutive activities takes a high degree of commitment.

In the selection of markets, we also played for the long term. For example, in Latin America we began our investments in 1988 at a time when high inflation, even hyper-inflation, debt restructuring and uncertain political prospects made the short term problematic. But, our investments were for the long term, and in 1993 we achieved overall profits and see continued good performance for the future.

Through commitment, we have been able to learn about and benefit from the value of early market entry and consistent follow-through. In both Australia and the U.K. we entered the market by acquiring paging business in 1987. We recognized that those businesses were limited, but we also believed that these markets would liberalize and present us with other opportunities. Although the U.K. did not work out for us, Australia did in a bigger way than anticipated. We have a valuable stake in OPTUS, the second general carrier and second public telephone operator in that country.

In Germany, we established our presence in 1988 and participated in a losing bid for the D-2 cellular license a year later. We remained active in Germany, however, and participated successfully in the later, winning bid for the E-1 license. We are now building the E-Plus business, and you will have the opportunity at next year's conference to enjoy DCS-1800 service from E-Plus.

Another aspect of BellSouth's commitment has been our ability to fully draw upon all of the corporate resources. I have already noted, both explicitly and implicitly, the financial resources that have been applied to our global development. In the final analysis, however, the most important indicator of commitment is the ability to draw upon the skills, experience and expertise of the people of the corporation.

Financial commitment can be scaled, measured and finally dealt with in some degree of precision. The deployment of people resources, on the other hand, invariably involves tough choices about opportunity costs and sometimes even tougher intangible factors. It takes real commitment for any company to deliver its people resources, and BellSouth has demonstrated this commitment.

In the planning and bid preparation leading up to OPTUS, we called upon more than 120 of our leading engineers, marketers and people from other key functions. In staffing OPTUS, we supplied 70 skilled employees for assignments of up to 3 years. By that time, we and our partners expect that OPTUS will have hired or developed its own skilled people.

I hope by this time you have a sense of the strategies that have worked for us and that we continue to employ. As I said at the beginning, they are not terribly different from those that most global participants are following. I suppose that differences at this stage of industry globalization are to be found principally in the quality of execution, and in that regard I hope that BellSouth comes out favorably in your independent evaluations.

Now is a good time to return to my earlier topic of the strategic importance to BellSouth, and to the industry, of supporting and fostering certain key principles of liberalization and competition.

BellSouth operates today in a wide range of regulatory environments, from New Zealand, where there is essentially no industry regulation, to the U.S., where the industry is pervasively and in many ways overly and dysfunctional regulated. There are also many markets where we would like to operate but cannot because liberalization and competition have not been accepted, or the principles governing entry and operation are inadequate.

Our strategy begins with a premise that liberalization and competition are good for both the markets and the players. No doubt this premise is not universally accepted, nor will it be universally seen as credible coming from a company which may be perceived as a diehard monopolist in its traditional market.

Let me offer a few examples in an attempt to deal with both issues:

• In Australia, Telecom Australia enjoys both a new name, Telstra, and the best financial performance in its history since the advent of competition through OPTUS. This has happened in spite of the fact that OPTUS has secured more than a 30 percent share of the cellular market and almost a 20 percent share of the international long distance market in its brief history. The market has been energized, to the benefit of both Telstra and OPTUS, and consumers now experience choice, innovation and markedly improved service from Telstra. Without doubt, the industry's image has been elevated.

There are numerous anecdotes and lessons in the competition in Australia. None is more telling than the fact that the market stimulation has been of such great value that Telstra offered, and OPTUS accepted, lower interconnect rates on cellular so that OPTUS would have even more leverage to expand its base of 300,000 customers.

• Across the Tasman Sea, New Zealand Telecom experiences consistenly improved performance at the same time that BellSouth New Zealand has established itself as a cellular competitor and Clear Communications has established itself as a viable long distance competitor.

• In the U.S., the combined market value of the companies of the old Bell System combined is nearly three times the market value of AT&T prior to divestiture. This is despite the fact that all have lost significant market share to competitors since divestiture.

• In Germany, in less than one year of cellular competition, the cellular market has almost tripled, with growth running at 80,000 - 100,000 new customers per month. Telekom is not only benefiting from its own cellular revenue growth but from substantial interconnect revenues as well.

• In the U.K.,BT continues to perform very well in all categories and has capitalized on its freedom to invest internationally. Again, this is despite very strong and successful competition from Mercury, Vodafone and a host of emerging players.

• And, again in the U.S., BellSouth, recognizing that the only way it and market will benefit in the long run, fully supports the principles of fair competition the need for players, old and new, to have the freedom to compete. This position is a matter of record in pending legislation and with the numerous regulatory agencies where we are regulated.

I will concede that each of the situations mentioned is complex, and has many facets deserving of more comprehensive analysis.

Nevertheless, the correlation between liberalization and competition, and benefits, seem inescapable.

Liberalization and competition in each of the markets mentioned are supported and fostered to varying degrees by policy makers, the new entrants and, over time, by the former monopolies themselves... at least in a managed and controlled way.

Managed and controlled... this is my second premise: that the best way to ensure the benefits of my first premise is for policies to be applied that manage and control the introduction and, until no longer needed, the conduct of competition.

Here, in Germany, such deliberations are now in progress. Also, in the Netherlands, France, Denmark and in fact the EU as a whole, the process is underway in contemplation of the implementation of the directive for 1998.

So, what are the key principles of liberalization and competition which I have suggested are worthy of the support of the industry? They include:

- First, establish in each market agencies that have the mandate to mediate disputes between carriers and to enforce clearly defined rules of entry and competition. Further, without a mandate to enforce, the mediation will become meaningless.

- Second, ensure that the institutional advantages of the monopoly carrier are sufficiently mitigated. These include the ability to cross-subsidize services and the ability to unfairly use market dominance.

- Third, provide interconnection at market-sensitive rates and flexible terms and, in addition, allow new carriers the ability to construct as much of their own required infrastructure as they deem to be in their best interest.

- Fourth and finally, insist that all carriers in matters of interconnection must deal with each other as equals, with mutual compensation and consideration being the measure of equality.

The continued advocacy of these four key principles is the final major strategy we have followed to establish ourselves as a global participant.

Once again, I invite both players and policy makers to consider support of these principles in your strategies for the continues globalization of the industry.

Directions in Technology and Costumer Needs at AT&T

John S. Mayo

Because information technology and customer needs are both global in scope and impact, my topic—"Directions in Technology and Customer Needs at AT&T"—should have broad applicability beyond AT&T and AT&T Bell Laboratories. I intend first to discuss trends in information technology. Then I will examine emerging service trends that are being driven by this technology and by evolving customer needs. These technology and service trends are creating the telecommunications revolution of the 1990s—which we refer to as networked multimedia communications. With such evolving multimedia communications, people will ultimately be able to get voice, data, and image communications anytime, anywhere and in any combination—and they will be able to do so interactively.

Almost all the trends in telecommunications, computing, consumer electronics and entertainment are being driven by exponential progress in just three technologies—the so-called "killer technologies" because they are so powerful.

So, let's first look at these key underlying information technologies that are expanding exponentially and thereby enabling and propelling the thrust of networked multimedia communications. Because they are also changing the work and home environments, these same technologies are driving customer needs. The more they can do, the more the customer wants. It has been an upward spiral that has lasted over three decades, and will surely last at least one or two decades more. What are these killer technologies? They are silicon chips, photonics and software.

Let's begin by examining the key information technology embodied in silicon chips—that is, microelectronics. Gordon Moore started the trend of measuring microelectronics progress toward increasing power by counting the number of components we can cram onto a silicon chip. As Figure 1 shows, the number is still doubling roughly every 18 months. However, there is evidence that this trend will have matured by the year 2010, with chips containing up to a few billion components each for highly packed circuits such as memory. For typical custom logic circuits, the range should be in the 10^8-to-10^9 region. The *actual* components per chip *produced* may be constrained by economics to be

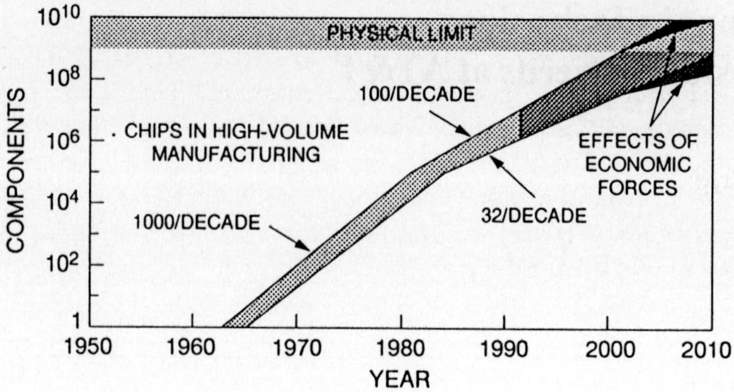

Figure 1 Components per Chip

somewhat fewer than the physical limit. The physical limit is set by 400 X 400 atoms being the smallest bulk-effect transistor that will work reliably at room temperature. We at Bell Labs are not backing off on bulk-effect devices, but there is evidence of a maturing of the technology.

The next frontier may be based on single-electron devices and ballistic transistors, whose behavior is described by the laws of quantum physics as applied to single particles. In a decade or two, such new devices may begin displacing some conventional devices, but that is far from a certainty, especially for room-temperature operation in low-cost applications.

A major application of microelectronics technology is the microcomputer systems that drive a lot of the systems opportunities. As the trend curves in Figure 2 show, microcomputer systems are doubling their processing power

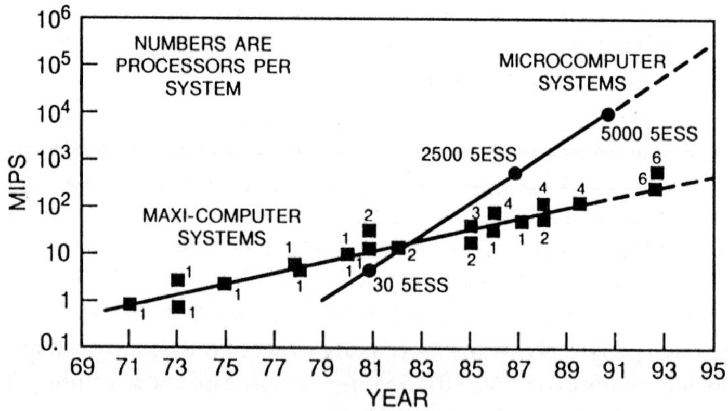

Data points show first commercial service for leading-edge systems.
Multi-microcomputer system points are for 5ESS® switch.

Figure 2 Trends in Processing Power of Computer Systems

every year, compared to every three years for mainframe or maxi-computer systems. Processing power is expressed here in terms of MIPS or Millions of Instructions Per Second. We see that microcomputer systems use ever-larger numbers of processors per system. The AT&T 5ESS® local switching system, for example, went from 2,500 microcomputers in 1987 to 5,000 microcomputers in 1991. And computing 20 years from now will still depend on the microcomputer, be it electronic or photonic, and on the distributed processing power of networks of microcomputers.

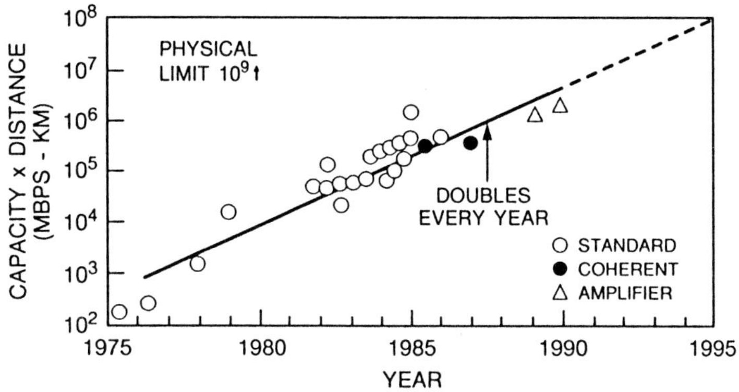

Figure 3 Lightwave Technology Capacity-Distance Achievements

Now, let's turn to the key information technology of photonics or lightwave. We measure the capability of lightwave transmission by the product of system bit rate and the distance that signals can travel before needing regeneration or amplification. As Figure 3 shows, that capability has been doubling every year for well over a decade, and can be expected to continue to double every year for at least the next decade before we reach known physical limits. That means we can expect a thousand-fold improvement over the capabilities of today's most advanced research lightwave systems. And today's practical in-service systems already transmit information at rates of up to 3.4 billion bits per second. That thousand-fold improvement would take us to trillions of bits per second—capacity that may be needed when we have massive amounts of high-quality, full-motion video and three-dimensional images.

Now, let's consider the pervasive and powerful information technology of software, which enables so many of the trends we have been examining. Today's major software problem is to increase quality and programmer productivity to meet exploding demands for software. As the trend curves in Figure 4 show, for some types of software, we are projecting a productivity range by 1995 of from 5,000 noncommentary source lines of code to over 50,000 lines per programmer per year. The increase in the upper range comes mainly from the growing reuse

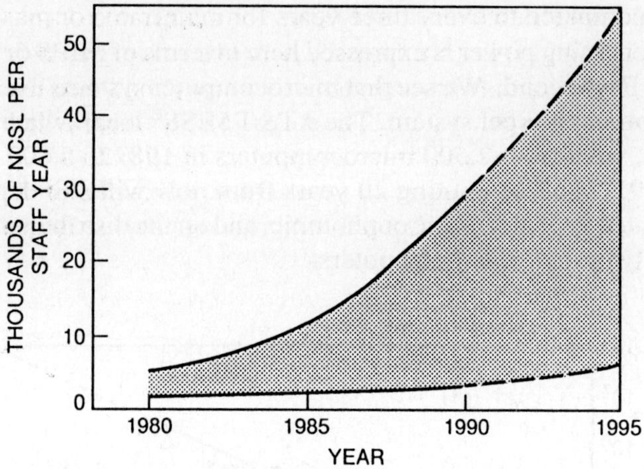

Figure 4 Productivity Range for Telecommunications Software

of previously developed and tested software modules. Without this approach based on reusable software assets, programmers produce one-tenth or fewer lines of code. In many software systems, for example, we find about 80% of code can come directly from a storehouse of reusable assets. Of the remaining 20% to be written at the line-of-code level, three-fourths of that can go into the storehouse and be reused in follow-on designs.

As we look two decades ahead, standards for software for electronic systems should be pervasive and widely deployed. And these approaches should further ease today's programmer productivity problem.

Thus, we see that the key underlying information technologies are incredibly rich and powerful. Now, let's consider what those technologies will do for telecommunications services. In short, they are propelling the digital thrust that is transforming our national telecommunications and information networks, as well as the way we work and play. That thrust—together with customer needs—is bringing about the merging of telecommunications, computing, consumer electronics and entertainment.

What is emerging is true multimedia communications—as I mentioned earlier—including multimedia products and services that are connected over telecommunications and information networks. And what will eventually emerge, as I suggested earlier, is the long-term goal of the telecommunications industry: the capability to provide voice, data and image communications anytime, anywhere and in any combination with convenience and economy. That capability will be provided primarily by highly intelligent photonics-based broadband networks of vast capacity.

Multimedia communications, as I said, is becoming the services revolution of the 1990s. First, let's define the term. A reasonable working definition is that multimedia refers to information that combines more than one medium, where the media can include speech, music, text, data, graphics, fax, image, video and animation. As multimedia systems evolve, these different media are being integrated—that is, they are linked, synchronized and commonly controlled, and are carried over a common integrated connection. The media can also be interactive or non-interactive. I am going to talk about a specific subset of multimedia: multimedia products and services that are networked; that is, connected over a telecommunications and information network.

Now, we are excited about multimedia because public switched networks can presently accommodate a wide array of networked multimedia communications, and the evolutionary direction of those networks will enable them to handle an increasingly vast range of such communications. Moreover, there is also a potentially vast market for multimedia hardware and supporting software.

Just how vast is the promise of multimedia? Market projections are difficult because it seems that no two people use precisely the same definition of the term. But Dataquest makes one of the broadest projections, basing it on the impact multimedia will have on six industries: video teleconferencing equipment suppliers; semiconductor manufacturers; workstation, software, desktop computer and related tools developers; consumer electronics companies; service bureaus; and other telecommunications and computer equipment companies. (The service-bureaus category includes revenues from local and long distance transport for multimedia information.)

Dataquest estimates the total worldwide market for multimedia products and services in 1996 to be $12.5 billion U.S. dollars, and it will leap to $95.8 billion by the year 2000. Of that number, nearly $27 billion will be from service bureaus' transport of multimedia information.

Apart from vast revenue potential, multimedia is also reshaping our traditional industries—including the merging I mentioned of key industries such as telecommunications, computing, consumer electronics and entertainment. For example, in various trials around the U.S., we are seeing the beginnings of a trend in video toward becoming both interactive and available on demand.

And between now and the early part of the next century, we will also see an evolving new class of multimedia information appliances, or information terminals, which will combine such functions as those of PC, video camera, phone, fax, copier and so on. (One early example, as you might know, is the AT&T EO Personal Communicators, which were made available in 1993.)

In particular, multimedia terminals will be ubiquitous in applications ranging from videotelephony and videoconferencing to interactive video and multimedia messaging. Moreover, these terminals will likely incorporate user-friendly capabilities such as speech recognition for ease of operation. There will be many opportunities both in networking and in terminal devices.

For example, AT&T has recently announced a new enhanced network service called AT&T PersonaLink℠ Services. Among other things, the service will enable consumers and business customers to send electronic "intelligent assistants" out on computer networks to find and filter information. These intelligent assistants are software programs attached to electronic messages. They can perform many of the time-consuming tasks that have discouraged more users from taking advantage of on-line services and the emerging electronic marketplace. The intelligent assistants are made possible by General Magic's new Telescript™ communications language. (AT&T is a founding member of a technology alliance with General Magic.)

And five major companies, including Sega and The ImagiNation Network, recently demonstrated video games for AT&T's Edge 16 modem product that will let remote users of Sega Genesis systems play games together while talking to each other simultaneously over standard phone lines.

Now, PersonaLink Services is one of the first examples of AT&T's approach to the interactive multimedia market. In this emerging market, AT&T wants to play the role of what we call "the missing industry." And that role is a "host" for a wide variety of digital content and multimedia applications developed by others. Customers will gain easy, timely and convenient access to personal communications, transactions, information services and entertainment via wired and wireless connections to telephones, handheld devices, computers and eventually television sets.

Independent sources for this digital content eventually will range from publishers to large movie studios to small cottage-industry software houses. Through platforms like PersonaLink Services, AT&T will accommodate applications developed and deployed by others, as illustrated by our initial offerings of on-line shopping and information services.

Since hosting is a function that connects end users to the content they seek, the topology of a hosting service must span the entire information infrastructure, from enabling content owners to package, test, and manage content, all the way to providing a personalized look and feel for the end user. Moreover, intelligent assistants can help customize services that suit the individual, whether a consumer or a business customer.

It is important to note that in the age of multimedia communications, consumers who are geographically separated from each other will not, for example, just play games together over networks—they will visit and find what is emotionally nourishing, and build their relationships. According to AT&T's vision, in this evolving age, people will seek new relationships based on telepresence, a new type of community, and a social experience independent of geography. This potential for interactive networks is quite unlike that found in the proposed availability in the U.S. of 500 pre-programmed TV channels on the CATV cable. The beauty is that consumers will have the freedom to choose any subject or service from the intelligent terminals in their homes. A key point is that they will be able to network clusters of friends or associates to enjoy such services as a group.

And helping to drive such trends is an enormous consumer appetite for entertainment, as demonstrated by today's strong trends in color TV, CATV, VCRs and camcorders—with projected strong trends in high-definition TV, or HDTV, and videophones. The American Electronics Association, for example, projects the worldwide consumer sales of HDTV receivers to be 4.3 million by the year 2000, and 30 million by the year 2010. One reason entertainment is so attractive to AT&T is that there is a lot of infrastructure that will have to go in place to deliver such services as HDTV and video on demand.

Now, I have said quite a bit about the multimedia impact on the consumer. But networked multimedia communications will dramatically change the nature of work, and will therefore have a broad impact on the business customer. Video teleconferencing, for example, is first coming into businesses to enhance productivity, save time, and reduce travel.

Consider the vision of multimedia telephony for business customers. In a few years, a person could be working with colleagues in branch offices in New York, Hong Kong, Paris, and Sydney. The combined task is to produce printed materials, presentation slides, and a videotape introducing a new product line. Each person is sitting at a workstation communicating with the other people using the AT&T network. Not only can each person see the others on a screen, but all can share a common work space on the screen. Using this shared work space, each person in the videoconference can retrieve text, images, and video clips from remote data bases, create material and merge it with the rest. Once finished, each participant can send the report, the presentation material with voice annotations, and the videos to others using electronic mail. Current developments in multimedia telephony are making this scenario more and more realistic.

Let's now look briefly at the broad range of networked multimedia applications that either exist today or are emerging.

We are seeing the emergence of multimedia telephony—of multimedia versions of two-way telecommunications. Examples include videotelephony, videoconferencing and collaborative work, in which people in remote locations collaborate on a common task. (I have just illustrated such capabilities.)

The projected markets are large, and include both business customers and consumers. According to Technology Futures, Inc., by the year 2000 the number of high-end videoconferencing systems will likely approach 400,000. For desktop video communications, the forecast is independent of whether the dominant platform will be a telephone device, a computer device or some integrated system. Shortly after the year 2000, the number of desktop units could climb to over 10 million, and it could surpass 60 million by the year 2010. Finally, the forecast for households using two-way video communications is for five million households by the year 2000, and by the year 2010 over 50 million households may have two-way video communications. By the year 2020, such communications could be as common as the telephone or television.

Another area of networked multimedia applications is real-time video on demand: a largely one-way form of communication in which video information is processed instantaneously and sent over the network. Here, for instance, consumers will be able to select any one of a large number of movies on a list of box-office hits, and order these selections at any time of the day or night that is convenient, in order to enjoy customized showings. Let me point out that market researchers such as Paul Kagan project pay-per-view revenue to be roughly $3 billion shortly after the year 2000.

Another example of real-time video on demand is interactive video—in which, for instance, a consumer might choose one of several possible angles from which to view a baseball game. A related form of such interactive video is interactive games, in which one consumer competes with another in various types of games.

In contrast to these examples of real-time video, multimedia messaging is essentially a store-and-forward application. Examples are text messages with voice annotations and multimedia electronic mail—examples of interest to both business users and consumers. For instance, a person could send someone a memo and include a video clip that gives that person some special instructions.

A rather broad area of networked multimedia applications is interactive information services. One example is home shopping with multimedia catalogs—in which, for instance, a picture of a product would be accompanied by a voice description and a video. Another interactive information service is electronic publishing with multimedia—illustrated by an encyclopedia that includes text, audio, images and animation. An article on, say, warblers would contain text describing the birds, images of each type of warbler, audio clips of the songs, and video clips showing their behavior.

Another broad area of networked multimedia applications is multimedia education and training. Here, the applications include multimedia procedure support, in which a multimedia approach will help someone perform a step-by-step procedure, such as repairing an automobile engine. And the applications extend to creating a virtual classroom, consisting of people in totally different places, united by means of videoconferencing links connected to personalized multimedia terminals. Moreover, classrooms and textbooks will be enhanced, and learning will be customized to the individual via access to remote expertise that will reside in multimedia expert-system data bases.

And further out—certainly by the year 2010—virtual reality will be an emerging consumer service, building on the increasingly widespread capabilities of interactive telecommunications and entertainment. Virtual reality will enable people to indirectly and remotely experience a place or an event in all dimensions. For example, a person will be able to "sample" a new home, a new office, or a vacation destination—without ever leaving home. The individual will be able to turn his or her head, for instance, and see different aspects of a particular scene—or be able to feel the texture of wallpaper from afar, by using an electronic glove. Virtual reality thus will greatly enhance the concept of telepresence. We will be able to visit from afar—and do it much more frequently than we could travel. And we will be able to select a school and to purchase a car, or virtually any other purchase that now requires examination in person.

But do not think of virtual reality as simply goggles and gloves. Think of it also as a flight simulator. A student pilot is put into a seemingly very real, but actually unreal environment. It is as if he or she were actually in an airborne plane. A similar kind of virtual reality is now rapidly developing in the field of laparoscopic surgery.

Now, let's sample the array of networked multimedia services that public switched networks can presently accommodate, and let's also look ahead at the potential services that network evolution should make possible. I will start with low bandwidth or low information-carrying capacity and move to high bandwidth.

At the low end, we have Plain Old Telephone Service or POTS. Here we have videotelephony—an area that AT&T experimented with as early as 1927 and introduced as the Picturephone in the 1960s. But today's technology makes it practical and available at reasonable cost. The most notable multimedia entry that's supported here at this low end is the AT&T Videophone 2500, the first color videophone that operates over conventional copper lines. To use conventional phone lines, both audio and video signals had to be compressed—the audio to 6.8 kilobits per second, and the video to the extremely low rate of 9.6 kilobits per second. The 2500 has a moderate frame rate of 2 to 10 frames per second, depending on the amount of motion in the picture. In addition, we have also introduced the Picasso Still-Image Phone, which sends high-quality color images over conventional phone lines.

Another multimedia possibility that is supported by POTS at the low end is integrated audio-image communications. One promising possibility here is the simultaneous ability to talk and send a fax.

The next higher level of network capacity is that associated with the Integrated Services Digital Network or ISDN—the international standard for integrating voice, data and images on the same line. At the capacity of two ISDN Basic Rate Interface channels, or 128 kilobits per second, we can have video that is compressed using the so-called p X 64 international standard for videotelephony, for people-to-people communications. This standard was designed for sending videotelephony over phone lines in multiples of 64 kilobits per second. Here we have the ISDN videophone, and that is likely to be a highly attractive form of videophone so long as copper wires make up the local telephone plant, as they do in the U.S. Such videophones have many applications. For instance, the AT&T Global Information Solutions (formerly NCR) automatic teller machine will have an ISDN videophone capability. Alternatively, 128 kilobits per second is also the capacity needed to support high-quality audio, including the possibility of stereo-quality sound over the telephone. And ISDN is already widely available for combined voice and data.

Beyond 128 kilobits per second, at the next higher level, we have the capacity of 1.5 megabits per second, or 1.5 million bits per second, what we call the DS1 rate in the U.S. This rate supports high-quality desktop video compressed with the maximum level of the p X 64 standard. The rate also supports the so-called Motion Picture Experts Group or MPEG standard that covers the interaction of people and machines. The DS1 rate supports excellent NTSC color television, though it falls short of broadcast quality.

Incidentally, AT&T's three-chip video-codec—or coder-decoder—set for video coding and compression can be used for ISDN video or for DS1 video. The chip set encodes video with either the MPEG standard or the p X 64 standard, and it decodes compressed video using either standard. This chip set will make it more cost-effective for manufacturers to bring advanced videotelephony and other multimedia features to desktop personal computers and workstations. We are talking about capabilities such as full-motion video, CD-quality audio and videoconferencing. The chip set also will help make it possible to combine full-motion video and audio in a broad range of interactive point-of-information and point-of-sale products—including interactive kiosks, cash registers and automated teller machines.

Now, the next higher level of network capacity is 45 megabits per second, the so-called DS3 rate in the U.S. This rate will readily support high-definition TV compressed to about 20 megabits per second, using variations of the MPEG-2 coding algorithm. That is a considerable reduction from the roughly one billion bits per second required for uncompressed high-definition TV.

The DS3 rate will also support near-studio-quality video that has been compressed by using variations of that same coding algorithm.

And finally, we have higher network data rates of 155 megabits per second and beyond. These rates will support many channels of high-quality real-time images and video, or highly specialized signals such as instant transmission of X-rays and CAT scans, or a series of real-time X-rays, remote views of surgery, or military applications such as satellite surveillance. Of course, the high bit rates will mostly be needed to carry the large numbers of multiplexed video signals that we will find in the heart of the network of the future—for example, in the trunks that connect big cities. In addition, the high bit rates will also be used to carry the high bandwidth within complex images, and will typically be required when there is a need for processing or manipulating the content of the images, especially multiple images. And some of the video signals in the network of the future will surely be associated with the ultimate multimedia service I mentioned earlier—virtual reality.

That, then, is a look at some of the networked multimedia services that public switched networks can presently accommodate, along with some of the potential services that network evolution will make possible. However, in the case of the very high-speed signals, the local telephone plant in the U.S. cannot yet support their distribution. That is an opportunity for the U.S. cable television industry as well.

In summary, the incredible richness and power of information technology are enabling and propelling the revolutionary thrust of networked multimedia communications. As I have illustrated, today's public switched networks can presently accommodate a wide array of networked multimedia communications. And the evolutionary direction of these networks will enable them to handle an increasingly vast range of such communications.

This revolutionary thrust is at hand, and will shape the way we live, work, play and travel as we enter the 21st century. And, most importantly, there is a convergence of ever-expanding customer needs and the rich information technology that will help satisfy those needs.

Tradition and Future of a Global Player

Rt. Hon. Lord Young of Graffham

Opening Remarks

I am the last speaker in what has been a very full two day Conference. Everyone has thanked and praised Munchner Kreis and I would like to add my small voice to the chorus. There can be no doubt that the founders of Munchner Kreis were both far sighted and persistent and have been very influential in shaping the future of telecommunications in Germany and through Germany the European Union. All European customers owe Professor Witte and Herr Schwarz-Schilling a debt of gratitude for the changes they have made in Germany.

Yesterday, the Titans or rather the Goliaths of today's industry spoke about how they might bestride the world beyond their own fortified fiefdoms. Today, some of the Davids have taken the floor. In Cable and Wireless' long history, we have been and still are both David and Goliath. The ability to be both David and Goliath and to create a new customer-focussed alliance by synthesising experiences is probably the most important characteristic of tomorrow's global players.

Introduction

The great German philosopher of history, Herder, once said that nations were like individuals and had special characteristics. If Herder is right about nations, his dictum is even more applicable to companies.

In the telecommunications world, there are noticeable differences between carriers whose origins were Governmental and whose history is essentially as a branch of Government - and these are the vast majority of carriers in the world - and carriers founded by private sector enterprise.

In the latter case, each company tends to reflect some dominant characteristics of its individual founder. Bill McGowan, the founder of MCI, was a street fighter prepared to go all the way to bring down the walls of Jericho to succeed. In addition to making MCI America's No.2, his battles with the Regulatory System ended with the break up of ATT; once ATT lost control of the local loop, MCI soared to become the formidable force it is today. The McGowan culture is still very strong in MCI - the all out effort to win customers by almost any means and to increase market share. US Sprint's origins in Southern Pacific are reflected in concentration on first class infrastructure.

The nationwide fibre network was built with great gusto and remains one of Sprint's success stories. In the United States, there are countless Mom and Pop local telephone companies all with countless quirks which differentiate them from one another and from the giant RBOC remnants of the old Bell system.

ATT itself retains the spirit of scientific discovery that so infused Alexander Graham Bell. Bell Labs may be huge but many an important innovation in the semi-conductor and opto-electronic areas has originated there. ATT's multiple investments in the small companies playing with critical elements of tomorrow's hardware and software are all part of this old spirit of scientific enterprise.

Tradition

The founder of Cable and Wireless was not a scientific discoverer like Alexander Graham Bell nor was he a software genius like Bill Gates or Steve Job. Sir John Pender was a plain old textile merchant with a sense of commercial adventure second to none. He was quick to see that a lack of real time information on the price of Virginian cotton was putting him at a competitive disadvantage. To rectify this information gap, he raised finance on the Stock Market to build the first cross Atlantic cable - in those days an event as awesome as Armstrong's moon landing at the end of the sixties. Although the cable failed, Sir John's spirit of enterprise did not fail. He collaborated in building a second Atlantic cable and went onto wire up the world - the Americas, Africa, Arabia, India, South East Asia and Shanghai.

Sir John Pender was probably the world's first global telecommunication player and yet he was not from the telecommunication industry - a fact we need to bear in mind when considering the potential global players of tomorrow.

The Cable companies Sir John founded were all pioneers in their time doing what many then thought impossible dreams or fantasies but integrated they gave Sir John and his successors a monopoly power that caused a strong reaction from ATT and the PTTs of Europe. The cable companies also ignored the threat posed by Marconi's discovery of wireless and in 1929 Cable and Wireless was established through the merger of Pender's cable companies and Marconi's wireless companies - possible the first convergence megamerger!

The merger came just in time. The 1930's saw a depression in world trade, a rise in competitive wireless telegraph services and the expansion of pure telephony. International facilities supplying seamless telegraphy for the business customer were not enough - a warning here for our modern Titans pursuing the Fortune 500! The company initiated a new policy of pursuing franchises to operate a broader range of telecommunications in many foreign countries. It was during this period that Cable and Wireless gained the franchises to run Bahrain's domestic and international services (in 1931) and Hong Kong's external wireless services (in 1938) - the two franchises which were relatively insignificant in 1939 were to provide the business launch pad for the company's return to the private sector in 1981 after being nationalised for 35 years immediately after the Second World War.

Although a former Cabinet Minister myself, I have to admit that Cable and Wireless benefited little from its period of public ownership. Indeed, its withdrawal from Latin America, where it had a dominant position, and its move into the maintenance and manufacture of equipment in the United States and Europe to maintain employment, look positively mistaken in hindsight - a warning for carriers contemplating diversification from their core business. Will the carriers paying over the odds to have a share of some of the US content providers gain anything in the long-term? BT moved in and out of Mitel quicker than C&W moving in and out of equipment maintenance and manufacture in Europe and USA.

The franchise business developed in the 1930's, although somewhat scattered and fragmented, proves to have been critical for carrying Cable and Wireless through its bleak period under Government control. In the 1930's, Hong Kong, Bahrain, and the Caribbean were unimportant start up businesses compared with the external services of UK, Australia, Canada, India and South Africa, but by 1981 they were the vital revenue earners that set C&W on a new path, as the first privatised telecommunication company of the modern era. Each franchise, however small, can be a processor in a network of distributed processors.

The privatisation of Cable and Wireless in 1981 proved to be thoroughly cathartic.

The first, and perhaps most important, consequence of the decision to privatise Cable and Wireless was to instil a new spirit of 'can do and must do' instead of the bureaucratic tendency to refer everything to committees. The management were galvanised by the urgent need to meet tight timetables. Problems, however difficult, were there to be solved and solved constructively, and for the first time in years - with an eye on the bottom line.

In successful privatisations, the leadership, management and key employees gain a sense of freedom and entrepreneurship to develop and implement a new corporate strategy.

In Cable and Wireless case it was a matter of returning to the spirit of its founder. My predecessor, Lord Sharp, in effect acted as if Pender reborn. He completed within 4 days - from start to finish - the arrangements to acquire control of the Hong Kong Telephone Company in 1984 - the largest acquisition in Cable and Wireless' history and the driving force behind the growth of the company even today. Such rapid and conclusive decision making would have been impossible if Cable and Wireless had been a public sector company or had retained the bureaucratic processers of an ex public sector company.

Many of you are familiar with the record of Cable and Wireless' activities during the 1980's. Some of the Goliaths here fought hard to check us; one helped us, as did some of the Davids here. Unlike every carrier here looking today to expand out of a well established domestic base, Cable and Wireless spent much of the 1980's restoring a domestic base - Mercury - from our distributed global operations. Under Lord Sharp, Cable and Wireless acted almost like Cassius Clay "I dance like a butterfly but sting like a bee". Just before he retired he wrote an article 'Who dares Wins', which provides a useful record of Cable and Wireless' post privatisation performance -

"Actions speak louder than words. Since 1981 Cable and Wireless has followed a multi-pronged strategy which would have been virtually impossible to conceive, let alone implement, under the coat-tails of HM Government:

- it has expanded its activities in the Pacific Basin, the world's greatest growth area
- it took over control of Hong Kong Telephone Company
- it has recently restructured its operations in Hong Kong merging both the national and international telecommunication operations and created Hong Kong Telecom, the largest company on the Stock Exchange of Hong Kong
- it has established joint ventures inside China - the only foreign company permitted to operate telecommunications inside China
- it has established Asiasat, a joint company with China and Hong Kong to provide the whole of China with its domestic satellite operations
- it became the first British company to be listed on the Tokyo Stock Exchange
- it has established a joint venture with Japanese partners (International Digital Communications) to provide an alternative international service in Japan: a move which is unprecedented in the history of telecommunications in advanced industrialised countries
- it has expanded its network in the United States and now has a coast to coast digital transmission system in the US where it operated in over 50 cities

- it has installed and is operating the UK end of PTAT, the first private sector submarine fibre optic cable to cross the Atlantic
- it is constructing the North Pacific Cable (NPC), the first private sector submarine fibre optic cable to cross the Pacific
- it is constructing a fibre optic cable which connects up key Pacific Rim countries: Hong Kong - Japan - Korea (HJK)

Combining Mercury Communications' all digital trunk network in the UK, PTAT, the US network, NPC, IDC and HJK, Cable and Wireless will be able to provide all the peoples of the Atlantic and Pacific regions with a choice of international digital facilities for the first time in history - the Global Digital Highway. Today, the company is poised to become everyone's second carrier around the world".

<u>Customer Choice</u>

The main achievement of these post-privatisation years was to create customer choice where there was none and to reinforce it in the United States. This was done in the teeth of severe opposition from monopoly carrier incumbents in the UK, Japan, Australia and to a lesser extent Sweden. The greatest pressure came when Cable and Wireless alone of all carriers responded to the FCC wish to see international facilities competition in both the Atlantic and Pacific Oceans. To a large extent, Cable and Wireless was in the vanguard of the liberalisation process which started in the United States, came to the UK and Japan and is still penetrating Europe and other parts of the world. True to its traditions of serving customers in their own territories as well as providing them with first class international connectivity, Cable and Wireless created customer choice by extending its operational presence within the Triad while at the same time building the links between the shores of Europe and the shores of Asia. In doing this, Cable and Wireless was also pressing at the frontiers of regulatory policies in the UK, United States, Japan, Sweden and Australia. Some of this experience was also used by the D2 Consortium here in Germany, where we have a small shareholding.

The Future

One thing being a long-standing global player teaches you is that you cannot rest on your laurels.

One of my chief tasks when taking over as Chairman was to prepare Cable and Wireless for the next stage of its journey into the future across a decade of revolutionary technological and commercial change.

We all talk of technology turning the world into a global village. We all talk of serving the customers but how many of us in the business world recognise the qualitative distinctiveness of each customer and each set of customers - something Goethe would have understood well. Telecommunications technologies may well help overcome time and space within the global village but it cannot eliminate the cultural and psychological differences abounding within the different neighbourhoods of the global village.

Call back and card services may be competitive and convenient but they cannot replace the quality of service and direct customer care a local operator can provide.

Cable and Wireless operates in over 55 countries and it is our view that there is no substitute to serving the customer within the customer's own neighbourhood whether that is the snows of Siberia or the sands of Arabia. It is only by being with the customer in his abode that a carrier, however global, can become that particular customer's global carrier.

Given our presence in the 5 continents of the World, we have gone through a process of geographical focusing to guide our continuing expansion of operational activities. Our priorities are around our hubs and clusters in Asia/Pacific, Europe, Caribbean and America. Within Asia/Pacific, we have three thrusts; Hong Kong, China and South East Asia; Japan and NE Asia; including East Russia; Australia and the Pacific islands. We have redefined Europe along lines General De Gaulle would have approved - from the Atlantic to the Urals, and here we also have 3 thrusts; Mercury and second forces in Germany, France, Italy; expansion of the Baltic Cluster around Latvia; Bylerussia, Petersburg/Moscow; collaborations with the smaller European PTTs. In the Caribbean, we are concentrating in the Caribbean rim countries such as Colombia and Central America; and in the United States we are following the rapid churn in the wake of the collapsing MFJ before we take the next step. We have secondary interests in Southern Africa and the Middle East which could become significant over time.

By establishing ourselves within the distinctive neighbourhoods of the global village, we plan to serve our multicultural and multilingual customers from within their own culture and language. This is quite normal practice for multinational companies in other sectors but is rare in the telecommunication sector where national monopolies bounded by national sovereignty have been the norm.

Cable and Wireless as the world's only flagless flag carrier has been the exception and will continue with its established policy of selective expansion of operating presences to serve customers on their doorsteps. By doing this we will also be extending the global network to some of the 4 billion people in the world without telephones.

This policy in effect plays in Cable and Wireless' traditional strengths but is as relevant for the future as it was for the past.

What about alliances? I can hear the question being raised. Everyone else is forging alliances - BT/MCI; DT/FT; US West/Time Warner; ATT/World Partners. Why not C&W?

The answer is relatively simple. Cable and Wireless is already an international alliance. We have more allies that any other carrier and our allies are more variegated than any other sets of allies. We are ready to work with any David or any Goliath in any part of the world if it makes local sense to both parties.

It is for consideration whether the global village can best be served by a few oligopolistic alliances between dominant carriers or whether local sensitivities and sentiments will prevail? The dominant carrier alliances are very recent. They have yet to make their mark outside their own territories. Even if they succeed, MacDonalds has not checked the success of local and regional foods; Pepsi and Coca Cola have not altered the localised drinking habits of the British, Germans, Chinese or Japanese.

But will not technology change all this? Cannot new services be longlined from one country to another without the customer even knowing? Will not new service providers connect onto the new superhighways anywhere in the world to meet customer demands anywhere else at any time of the customer's choosing? Will not geographic presence become irrelevant?

The answers are Yes and No, Technology will make a difference. It will put more power in the palm of the customer's hand and fundamentally alter the balance of power between the customer, carrier and supplier.

ATM (Asynchronous Transfer Mode) will alter the way costs and benefits are distributed in the telecommunications value chain; it will shift the balance of power away from the giant PTOs and their suppliers towards end users and the vendors of customer networks. The days of customer and particularly business customer liberation are at hand.

ATM gives customers the opportunity to redraw the boundary line between themselves and the PTO and to enhance the functionality within their own networks. It promises customers universal interconnection for all services and simplified network management. The immediate benefits of ATM are already being enjoyed by companies operating their own networks across multiple sites with a need for high bandwidth communication and multimedia applications.

While ATM will help liberate the business customer, PDAs (Personal Digital Assistants) will do the same for the individual customer.

In this new environment, customers are unlikely to accept standardised solutions worked out by PTO alliances. The customer will be looking for special customised solutions to fit his or her specific wants - something recognised by General Magic in California. The carrier both physically closest to the customer and capable of providing these customised solutions is likely to become that customer's carrier of choice.

As an international alliance with multiple points of presence, Cable and Wireless welcomes the liberation of the customer from dominant PTO dependency and seeks to partner the customer in providing access to all points of presence in ways determined by the customer's wants. Cable and Wireless' points of presence become more valuable as customer's use new technology to provide their own solutions around the world.

Cable and Wireless might not have its own Martlesham, Musashino, Bell Labs, Darmstadt or Rouen, but its engineers have their eyes and ears open all the time and we are not afraid of deploying state of the art technology whatever its source. We were the first to create an all digital network in Europe - Mercury. Our PTAT and NPC cables were the largest capacity cables in their day. We were the first to introduce PCN or DS 1800 in the world. We will be among the first to introduce VOD when we do so in Hong Kong later this year.

Alongside ATM and broadband technologies, the new wireless technologies are equally liberating. GSM in Europe and around the world is creating a whole new network in which customers can cross frontiers with one phone calling anyone anywhere anytime independent of the fixed networks.

It is my belief that wireless will not only give customers greater choice where they already have universal service but will also accelerate the spread of telephony to the vast majority of mankind without a telephone. The wireless phone is becoming smaller and lighter while increasing its functionality all the time. There is no reason to think that over the next decade it will not become like the calculator and transistor - something that will provide the majority of mankind with access to the global network.

We have been reviewing the technologies of convergence. We have been impressed by the important innovatory role of the small companies in California and some of the consumer electronic companies in Japan. We are establishing a network of expertise to plug into these and other centres of excellence. At the same time, Mercury itself has launched an imaginative programme looking for some killer applications.

Should we make any breakthroughs, the application will be developed not only for Mercury but redeveloped for the distinctive needs of our multilocalities or perhaps multi customer windows on the world.

Conclusions

The future is full of promise for customers above all. Carriers will have to be multilocal and fleet of foot if they are to meet the requirements of tomorrow's customers. Global players will have to be like Mercury, the winged Messenger of the gods, if they are to survive into the 21st century.

Future Development

Summary Record of the Panel Discussion Chaired by Mr. Irmer

Theodor Irmer

Making the connection from this second day of the Congress to the final panel session, ***Mr. Irmer*** referred to ***Mr. Mayo's*** presentation which provided a lucid overview on the power and richness of future technologies, leading to almost unrestricted possibilities. He raised the question whether there was not a danger that this seemingly unlimited growth would sooner or later result in overcapacities which would not be used and, therefore, uneconomic. As ***Mr. Mayo*** demonstrated, the capabilities of technology increase rapidly and at the same time rates decrease, driven by competition; however, innovation and consequently depreciating cycles are becoming shorter and shorter. The question arises whether this evolution might end in a downward spirale, leaving network capacities idle and, consequently, result in an economic turmoil?

Lord Young noted that talking about overcapacities is an old story, but so far the predicted negative consequences never became true. Two major forces will in future balance the rapid growth of capacities: Firstly, multimedia will absorb, as a "bandwidth-hungry" service offering, many of the capacities and secondly, falling rates will generate more traffic. Looking at early history, the one-minute rate of an intercontinental call was about US $ 300.- (with today's purchasing power), whereas modern technology has today decreased this rate to some 50 cents; at the same time, it would nevertheless still generate a growth of traffic on the inter-continental routes. ***Mr. Leive*** supported this view; it is his opinion that demand is still more increasing than capacities are growing, and one should have confidence - based on experience - that this trend will also continue in future.

Against this scenario, ***Mr. Irmer*** invited comments from the regulators' point of view; how would they feel being confronted with the technical evolution as described in ***Mr. Mayo's*** presentation?

Mr. Carsberg did not see this evolution as a national regulators' nightmare; he felt that the problems will be more in the area of the global rather than in the national regulation. The national regulators will continue to promote competition and to protect customers; their lives would become even easier as there would be more choices to make. However, they would certainly have to supervise more closely that strong distribution powers would not at the same time be strong powers for the contents of information they are distributing, and vice versa. Here open access would have to be secured as well as common tariffs for the carriers and a non-discriminatory pricing system, regardless of volumes but equal for similarly placed people. It will not be easy to achieve such an openess; even today, with e.g. "video-nasties" of pornographic contents, national regulation of contents is difficult to achieve, and will be even more though in the future. As far as national regulation on infrastructure is concerned, ***Mr. Carsberg*** noted that competition on the infrastructure would be welcome as it would advance contests of technology; one might eventually find out which technolgy would be winning this contest.

Turning to ***Lord Young***, whose company has been a genuine global player since a long time, ***Mr. Irmer*** enquired whether he would see a need for global regulations, and more specifically in the area of global standardization.

In his reply, ***Lord Young*** expressed his overall satisfaction with the existing standardization system; for example: GSM, although in his opinion overspecified to some extent, is about to become a global standard. But in general, fewer - and not more - standards should be developed which should then remain stable for quite some time. In any case, it should be left to the market forces to shape the global standardization process. As a negative example, ***Lord Young*** quoted the recent European attempt to develop an analogue technology-based HDTV standard as a dead-end initiative. Standards should certainly not be set up as a means of protection as it happened often in the past, for example in Europe and the United States of America.

Mr. Eidenberg concurred broadly with Lord Young's point of view and he warned that standardization should not go too far. Disputing some flaws in standards should not be understood as if there were no need for ITU's global standardization. As far as regulation is concerned, it should mainly serve the purpose of mediation and fair competition, rather than dealing with detailed problems like access to the market, market applications, functionality of services and selection of technologies. That kind of regulation is in his opinion outdated, and countries who have not yet set up their regulatory regime should learn a lesson from this experience. ***Mr. Leive*** added that the future role of the regulator should mainly be that of assuring universal access to information, similar to that of the past when the regulator had to define the universal services. Concerning standardization he noted that the role of governments would decrease, resulting in a decentralized standard-setting process, largely determined by the private telecommunication sector.

Looking towards the future, ***Mr. Mayo*** stressed that priorities in global standardization will increasingly be set by the global customer which will lead to quite some pressure on the standardization organizations. Moreover, the so-far separate standardization processes in telecommunication, computer industry, consumer electronics and entertainment industries will have to come together, challenging the established standardization procedures and organizations. ***Mr. Irmer*** mentioned that the many Fora (such as ATM Forum, TMN Forum, etc.) which came up recently might serve as an interesting model of how such overall standardization might be done in future, although quite a number of questions would have to be resolved until such "open standardization" in whichever form would come into being. At this point, ***Lord Young*** remarked that governmental agreement on global standards would in future be almost impossible, only to a small extent - if at all - concurring with a similar remark made earlier by ***Mr. Leive***.

Mr. von Pattay recalled the discussions from the foregoing day on the subject of why global players are going global (e.g. because of cost cutting, market shares, global customers), and he was wondering whether this might eventually lead to problems within a country (i.e. in one and the same country, different "conglomerates" with different technologies, infrastructure, etc. would exist, thus creating interworking problems of all kinds). In such a scenario, should one leave these problems to be solved by technical means ("universal, all-adapting chips") or would it be for the national regulator to restore the "national harmony" by on-time regulations?

Mr. Carsberg's position on this issue was very clear: every measure which would make interworking difficult would be interpreted as an anti-competitive practice to gain a dominant position. To safeguard open access would be the job of the pro-competitive national regulator.

Along these lines, Mr. ***Eidenberg*** underlined that multinational customers, ahead of other customers, would not allow to become "prisoners" of national proprietary standards and policies; such attempts would be a loosing strategy as experience in the computer market demonstrated. ***Mr. Mayo*** agreed and requested that the customer should find equal conditions everywhere, and if they do not exist from the outset they will have to be made available.

While starting to summarize the results of this panel discussion, ***Mr. Irmer*** pointed out, that this Congress expected by no means to receive verified and proven strategies of the global players. The objective was to learn from each other, to exchange experiences and to discuss possible future strategies. Although the main road seems to be largely the same for all players, marked by milestones such as liberalization, privatization, competition and globalization, there seem to be nevertheless some differences; e.g. when comparing Europe and the United States of America. While each global player has its "global player toolkit", the European tool kits (notably with the exclusion of the United Kingdom) are new, perfect, shiny and ready for use, whereas the USA tool kits are more like a handbag. Their tools are no longer so new and shiny, even a little bit worn out because they have already been used heavily. No doubt, there is no shortage of good ideas, plans and objectives in Europe but their conversion into clear market strategies seems to be a good way ahead. For example, re-balancing of tariffs is one prerequisite for "going global" but in most European countries there are only cautious attempts, watchfully supervised by governments - and this holds true also for other issues. The year 1998 will come up soon, and there is not much time left to make European telecommunication companies fit and firm for the rather difficult situation they will then be facing. It appears that many future global players in Europe are somewhat handicapped by their former role as monopolies: to do everything themselves in an all-embracing fashion and to do all as perfect as possible. In contrast to the other side of the Atlantic, the global players seem to be more pragmatic. They are driven by economic considerations - they simply worry less about what might possibly happen in, say, 5 years but they take their chances now. For example, they started going global with good old POTS, adding new services when appropriate whereas "new services" (which ones?) are a major part in the European global strategies. ***Mr. Irmer*** concluded that such considerations would be his personal subjective opinion but he felt that the European global players should nevertheless learn from the US experience - in their own interest, as time is running short.

Mr. Claus felt that such views would not reflect correctly existing reality; in fact, there have been many changes and the new European telecommunicaton companies ("Telecoms") would be much different from the former monopoly PTTs. He pointed out obstacles which these new companies are facing. For example, in Germany the re-balancing of tariffs started but they are being watered down by regulatory, political and other influences for which the German Telekom cannot be held liable. As another example, ***Mr. Claus*** referred to the discussion currently underway in the United States of America on setting up super information highways, whereas in some 40 cities in Germany switched 140 Mbit/s services are available since 1988 already. To point to the USA as the master in telecommunication would not be appropriate; also the European upcoming global players are willing and prepared to face the challenges ahead in their future role.

Mr. Irmer admitted that his summary remarks were intentionally somewhat provocative and stated that this last frank and open controversial discussion has served its purpose by highlighting differences in opinions and positions which is the objective of any panel discussion. It is self-evident that such different viewpoints exist; nobody can predict what will be the situation in the telecommunication market place in, say, 5 or 10 years. One will find out at that time which past opinions and positions were right or wrong. ***Mr. Irmer*** thanked the panelists, the speakers and the audience and terminated this second panel session of the Congress.

In his closing remark, ***Mr. Witte*** pointed out that, although his remark would be final, nothing else that was reported in the course of this Congress, should be considered as being final. As presentations and discussions during this Congress so convincingly demonstrated, "going global" has just about started to unfold its dynamics. During the two days of the Congress, all participants learned a lot about what is going on in the world of global networking. Much information was given on what global players are planning; consequently, one will have to analyze and study such plans in order to evaluate positive and negative effects. Last but not least, this Congress also presented an opportunity to personally meet the people who are behind the global players. This Congress provided indeed much food for thought, and everybody is invited to help himself (or herself).

With cordial thanks to all speakers, panelists, the audience and all who helped behind the curtains to turn this event into a success, ***Mr. Witte*** closed the last session of this two-day Congress on "Global Players".

Index of Authors

Edgar L. **Brown**, Jr.
President of
Bell Atlantic International
1310 N. Court House Road
5th Floor

USA-Arlington, VA 22201

Helmut **Ricke**
Vorsitzender des Vorstandes der
DBP Telekom
Postfach 20 00

D-53105 Bonn

Charles B. **Coe**
President of
Bell South International
1100 Peachtree Street, N.E.
Suite 400

USA-Atlanta, Georgia 30309

Marcel **Roulet**
Président de
France Télécom
6, Place d'Alleray

F-75505 Paris Cedex 15

Eugene **Eidenberg**
Executive Vice President of
MCI Communications Corporation
1801 Pennsylvania Avenue, N.W.

USA-Washington, DC 20006

Dr. Pekka **Tarjanne**
Secretary General of
International Telecommunication Union
Place des Nations

CH-1211 Genf 20

David M. **Leive**
Latham & Watkins
1001 Pennsylvania Avenue, N.W.
Suite 1300

USA-Washington, 20004-2505

Sir Iain D.T. **Vallance**
Chairman of
British Telecommunications plc.
81 Newgate Street

GB-London EC1A 7AJ

Dr. John S. **Mayo**
President of
AT&T Bell Laboratories
P.O.Box 636
600 Mountain Avenue

USA-Murray Hill, NJ 07974

Viesturs **Vucins**
President of
Unisource NV
P.O.Box 20 42

NL-2130 GE Hoofddorp

Phil **Walker**
Vice President of
Sprint International
12490 Sunrise Valley Drive

USA-Reston, Virginia 22096

The Rt.Hon. Lord **Young** of Graffham
Executive Chairman of
Cable & Wireless plc.
124 Theobalds Road

GB-London WC1X 8RX

Prof.Dr.Dres.h.c. Eberhard **Witte**
Vorsitzender des Vorstands
MÜNCHNER KREIS
Tal 16

D-80331 München

Session Chairmen

Sir Bryan **Carsberg**
Director General of
Fair Trading
Field House
15-25 Bream's Building

GB-London EC4A 1PR

Dr. Theodor **Irmer**
Director of TSB
International Telecommunication Union
Place des Nations

CH-1211 Genf 20

Felix **Rosenberg**
Generaldirektor der
Schweizerischen PTT
Viktoriastr. 21

CH-3030 Bern

Prof.Dr.Dres.h.c. Eberhard **Witte**
Vorsitzender des Vorstandes
MÜNCHNER KREIS
Tal 16

D-80331 München

Printing: COLOR-DRUCK DORFI GmbH, Berlin
Binding: Buchbinderei Lüderitz & Bauer, Berlin